"Pyle shows himself here to be a cunning essayist who is able to bring to life a region little known. The layout of the book is poetic in itself—set in four sections that symbolically represent summer, fall, winter, and spring, each section in turn contains four essays—revealing a cohesion of the author's spirit with the seasons. . . . [A] love song to an overlooked—and overworked—land."
—*KIRKUS*

"*Wintergreen* constitutes a biogeography of the naturalist himself . . . The clear, complex topography of his language pays homage at every turn to the land he describes."
—*CHRISTOPHER CAMUTO, SIERRA MAGAZINE*

"In the southwest corner of Washington State lie the Willapa Hills, a temperate, rain-drenched land of perennial greenness. Still the habitat of fungi, mosses, lichens, and ferns, they were once the home of 'one of the greatest forests on earth.' But as Pyle so articulately states, years of improvident lumbering practices and economic greed have despoiled the hills, decimated the wildlife, and rendered the future uncertain. Out of this, his chosen home, Pyle has created a collection of vividly responsive observations and speculations about the diversity and requirements of life, from butterflies to bears."
—*LIBRARY JOURNAL*

"Pyle looks past the stumps and clear cuts to rediscover the essence of the Willapa Hills. Even though scarred from decades of logging this terrain is still a wilderness. The author puts the pieces together to create a vision of nature trying to survive our species' demands."
—*KRIST NOVOSELIĆ*

"A rare and wonderful book."
—*THE SEATTLE TIMES*

# Wintergreen

## Rambles in a Ravaged Land

Originally published by Scribner in 1986

Published by Pharos Editions in 2015



Library of Congress Cataloging-in-Publication Data

Pyle, Robert Michael.
Wintergreen : rambles in a ravaged land / Robert Michael Pyle
p. cm.
Originally published: New York : Scribner, ©1986.
Natural history—Washington (State)—Willapa Hills. 2. Ecology—Washington (State)—Willapa Hills. 3. Logging—Environmental aspects—Washington (State)—Willapa Hills. 4. Willapa Hills (Wash.) I. Title.
QH105.W2 P95 2001
508.797'92—dc21 2001031379
ISBN 978-1-94043-623-4

Cover and interior design by Faceout Studio

Interior map: Catherine Macdonald

Pharos Editions, an imprint of Counterpoint
2560 Ninth Street, Suite 318
Berkeley, CA 94710
counterpointpress.com

SELECTED AND INTRODUCED BY
DAVID GUTERSON

# WINTERGREEN

## RAMBLES IN A RAVAGED LAND

by
ROBERT MICHAEL PYLE

PHAROS EDITIONS | AN IMPRINT OF COUNTERPOINT | BERKELEY

# INTRODUCTION BY DAVID GUTERSON

*Wintergreen: Rambles in a Ravaged Land* appeared in 1986—the year of Chernobyl, and a year after the discovery of a gaping ozone hole over Antarctica. A consummate work of natural history, it announced the arrival of a stylist with élan in a field where currents of castigation and warning sometimes give "nature writing" the tincture of sermonology. *Wintergreen* felt personable, even optimistic. It had no truck with doom and gloom. Its author was possessed—it seemed by nature—of a robust and general enthusiasm, and of a buoyant appreciation for the fact of living. It was easy to imagine Robert Michael Pyle cheerfully rambling in his ravaged land, the Willapa Hills of southwestern Washington.

Imagine this I did, in 1986, while devouring *Wintergreen* cover-to-cover, a binge spurred by affinity and affirmation, or at least by recognition and identification, because here, at last, was a book about home, home as I knew it in my feet, so to speak, home as I loved and understood it. Other readers had described Pyle's Willapa as a metaphor for global environmental duress—or for something that went unnamed and felt vague—but for me, everything there stood for itself. In fact it was precisely the book's concrete bona fides—its familiar trilliums, skunk cabbage, and clearcuts—that made reading it feel so companionable.

*Wintergreen* took a place on my shelf alongside Robert L. Wood's *Olympic Mountains Trail Guide* and my *Climber's Guide to the Olympic Mountains.* I gave Wood's guide to a young, lost rambler, the climber's guide got chewed by voles, and *Wintergreen* went permanently borrowed. All three books have since been replaced, as has a fourth that, like the others, I pack regularly—Daniel Mathews' *Cascade-Olympic Natural History*, a supremely accomplished trailside reference that is out of print but shouldn't be.

Mathews' book appeared in 1988, the same year *Wintergreen* was published in paperback under the subtitle *Listening to the Land's Heart. Listening to the Land's Heart* is no less alliterative then *Rambles in a Ravaged Land*, but it strikes a more self-consciously poetic note, perilously so—in my opinion—suggesting as it does a preciousness about nature that is entirely absent from the text. *Wintergreen* is, in its way, poetic, but its poetry is constructed of the raw and tangible, and Pyle, lacking the Romantic impulse, is assiduous in its pages to avoid animism. His Willapa is meaningful to the extent that it is real. Its actual waters run downhill. "Heaven is here," Pyle has asserted, "angels are butterflies and bats, and the great beyond is the holy compost pile of the ages."

I came of age in Seattle in the mid-Seventies, an era when it sagged on the bleak edge of null, portal to the North Pacific. "Seattle in the seventies was the nadir of just everything," the superlative essayist Charles D'Ambrosio has written, and as his contemporary on that peculiar scene, I can attest that he's hyperbolizing accurately. D'Ambrosio's answer to the provincial doldrums was to cultivate an inner expatriation and to import his enthusiasms from more sophisticated locales beyond the pale of our rain. Trapped in the veritable outback of nowhere and feeling decidedly marginal, he became, he writes, "clever and scoffing, ironic, deracinated, cold and quick to despise."

Young white guys in the Seattle of the mid-Seventies, given that they had the luxury, plied one of two pervasive personas—either D'Ambrosio's Budding Expatriate, or my choice, Wilderness Romantic. I took to the hills whenever possible and eventually expatriated to a logging town where I went to work for the Forest Service in the spirit of Norman Maclean. This was during the heyday of clear-cutting, when mills ran 24 hours a day all over western Washington. I lived in a van, bathed occasionally in a river, read Gary Snyder, Edward Abbey, and John Muir, and while all of this was fundamentally delusional, it did get me rambling in a ravaged land.

For the Forest Service I toiled at "brush disposal," which meant cleaning up, in nominal and egregious fashion, after gyppo loggers with contracts. They felled, bucked, dragged, loaded, and drove off with whole forests; we burned what they left behind, on the theory that a modulated conflagration was optimum for "reprod." Our project never flagged. We lit up the sky on summer nights, adding scorched acreage to a running tally and sucking noxious smoke with pride. I fell so thoroughly in love with this folly that before long the Forest Service made me a crew boss. In clearcuts I was steeped, then. On slopes I cut my teeth. Eventually my initiation, with its ancient innocence motif, took me by its roundabout path into once, twice, and thrice logged hills, not the sterile hills inside my head but the living hills that, a decade later, I found again in *Wintergreen*.

There is a literature of regional affinity, and while its readership is by definition local, its value transcends its raw material. At its best it has a tempered, celebratory quality, or forwards a critique in tones of endearment, or with sad fondness acknowledges the constraints that bind us to provincial lives. Where changes are too much for locals to abide, it provides them with the balm of elegy, and gives what feels like testimony to their losses. I am thinking, here, in canonical terms, of Eudora Welty, Ernest Gaines, Edith Wharton, Hamlin Garland, Willa Cather, and Mari Sandoz—of all those quintessentially American fiction writers whose rootedness in place confirms our sense of home—and of nature writers like Wendell Berry, Terry Tempest Williams, Scott Russell Sanders, and Pyle, who mine home and place with a reach profoundly vertical (as opposed to the horizontal expansiveness of, for example, John Muir, Diane Ackerman, and Peter Matthiessen).

*The good God is in the detail*—apocryphal Flaubert—is a maxim understood by those nature writers who employ specificity to incite appreciation. A further and perhaps less noted function of the close rendering of the world's raw facts is the sense it can elicit of a private affirmation—like mine on discovering *Wintergreen*—the concordance one feels in the presence of fellow travelers sharing one's particular way-station. The action of this might be peripheral or corollary, but nature writers of the vertical variety nevertheless remind us that we're not alone—that we share, with others, resonating spheres, internal and external worlds.

Art's purpose, Tolstoy said, is to transfer feeling from one heart to another, or, as David Foster Wallace put it, "writing, at its best, is a bridge constructed across the abyss of human loneliness." At the bottom of that abyss looms the three-headed Hydra of distance from "the other," home, and nature. *Wintergreen* expressly addresses nature, but implicit in its pages is an address, too, of this tripartite and disabling loneliness.

Prior to *Wintergreen*, Pyle had published six books, all of them about butterflies. Subsequent to it he has published eleven more, including *The Thunder Tree: Lessons from an Urban Wildland, Sky Time in Gray's River: Living for Keeps in a Forgotten Place*, and, most recently, the poetry collection *Evolution of the Genus Iris.* There have also been more butterfly books, most pressing ambitiously beyond the early field guides, like *Nabokov's Butterflies,* which Pyle edited with Brian Boyd. The body of work coheres in the sense that the voice on the page remains affable and the authorial sensibility steadfast; there isn't much inflection over time. *Wintergreen*, though, is essential and foundational—the root text from which the rest have sprung.

The structural program of *Wintergreen* is symmetrical—four sections of four essays each, in total sixteen, a balanced and functional square number. At its midpoint it shifts from biogeography to reasoned disputation with human behavior—corporate logging in particular—before making way for the poetic impulses, local musings, and reflective metaphysics that define its concluding quartet. *Wintergreen's* itinerary, which lends direction to its journey, doesn't preclude rambling in the better sense of the word—rambling not as prolix meanderings, excess, or randomness of tone, but as the freedom Robert Frost described when, while discussing rhyme, he coined "moving easy in harness."

*Wintergreen* is unusual within its genre for its willingness to challenge the dichotomy between nature pristine and man the destroyer. "The beauties and biological interest of the logged-off landscape," its prologue announces, are its raw material—in other words, Willapa as it is, devastated but rife with fascinations. *Caveat emptor*, again from Pyle's prologue: "any attempt to recruit portions of *Wintergreen* in favor of regional timber-stripping practices will be *ipso facto* misquoting the book and taking the work out of context." Indeed, the sack of Willapa receives unstinting treatment from Pyle, who sees beauty everywhere—even in stumps—but is no apologist for timber companies, and names names without hesitation.

Pyle was surprised when Robert Finch and John Elder, while putting together the *Norton Book of Nature Writing*, selected from among *Wintergreen's* sixteen chapters "And the Coyote Will Lift a Leg." Here Pyle waxes philosophical and muses candidly on those metaphysical matters that inevitably invade—or pervade—Willapa. "A short-term optimist," he's "willing to conspire with the physics of fate (chance, really) to harvest luck from happenstance." He describes pessimism as "its own punishment, since it vitiates the will and makes one a pawn of circumstance," and immortality as "a permanent vacation from personality." In Pyle's view, we must work toward a vision "beyond heaven and humanism" if we are to put ourselves in an honest cosmic context and find equanimity in the face of our condition. After all, our transitory habitation of earth is in fields where "nature will not be eliminated," and where—as in the chapter title aphorism—"when the last man takes to his grave, there will be a coyote on hand to lift his leg over the marker."

It's been said—too often—that death and taxes are the only certainties in life, but in the Willapa Hills, it's death, taxes, and rain. For the sanguine Pyle, having long ago made his quietus with mortality, taxes are easily the most onerous of the three, and rain easily the most enduring. In fact, *Wintergreen's* abiding image is of healing rain, of rain repairing Willapa's wounds and reconciling its inhabitants to nature. Our eternal forecast, Pyle reminds us, is for rain both inundating and ameliorating, which is why we need, and will continue to need, this singular, celebratory, and heartfelt book.

— DAVID GUTERSON

# Wintergreen

## Rambles in a Ravaged Land

Then let not winter's ragged hand deface
In thee sweet summer, ere thou be distill'd:
Make sweet some vial; treasure thou some place
With beauty's treasure ere it be self-kill'd.

—WILLIAM SHAKESPEARE, *Sonnet 6*

The lark, the bird of light is there
in the bitter short days. Put the lark
then for winter, a sip of hope,
a certainty of summer.

—RICHARD JEFFRIES, *"Out of Doors in February"*

No one winterbook—no book—can find nearly
all that should be said of the West, the Wests.

—IVAN DOIG, *Winter Brothers*

# CONTENTS

For Thea

and to the memory of five great
Washington naturalists,
Daniel E. Stuntz, Frank Richardson,
C. Leo Hitchcock, Melville H. Hatch,
and Victor B. Scheffer

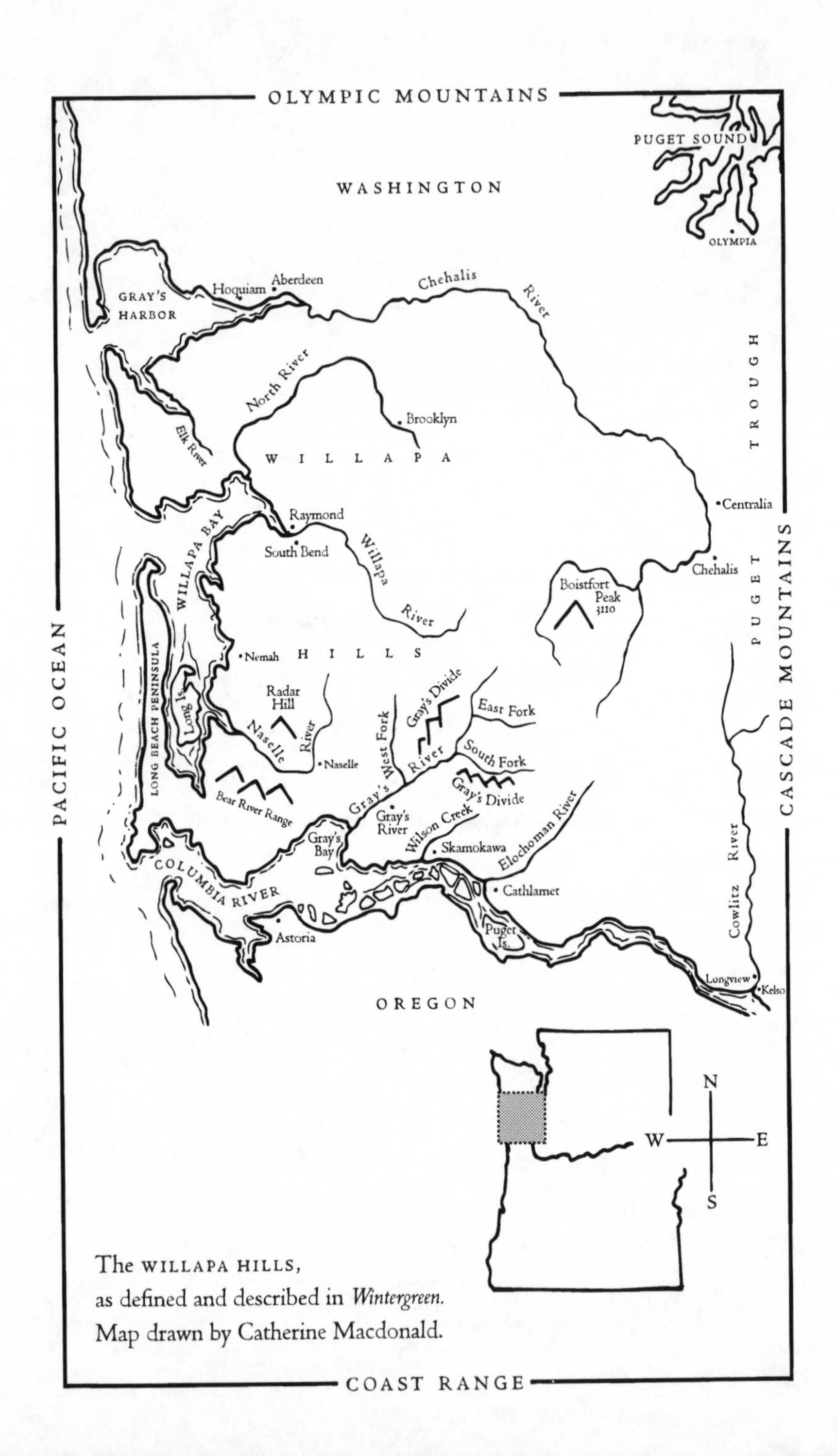


The WILLAPA HILLS,
as defined and described in *Wintergreen.*
Map drawn by Catherine Macdonald.

# Prologue

*An Evergreen and Pleasant Land*

AT ANY TIME OF the year and in any weather, my bedroom window frames a green and pleasant country scene. Halfway open, it makes a Kodachrome slide of the bucolic valley below, bordered by white sashes and molding. Timbered hills tumble down to a floodplain pasture valley, bounded below by a limpid river, itself spanned by an old gray covered bridge. Holstein cattle spot the meadows with black and white, and for half the year swallows pock and streak the broad skies.

I came to rural Wahkiakum County, Washington, at the end of the 1970s. This old homestead, known as Swede Park, represented release from the stress and distraction of the city. Gray's River, the stream and the town, seemed to provide the peaceful setting I needed in order to write full-time. The surrounding Willapa Hills, little studied by biologists, offered a fruitful field of exploration for a too-long-urban naturalist.

For the first few years, I maintained a pattern of commuting to Cambridge, England, for work that supported my writing habit when I was at home. By the time I settled in at Swede Park for good, I had lived in Great Britain four of the past ten years. So similar seemed the British Isles and the Pacific Northwest in some respects that I felt continually disoriented at first. To be sure, the differences are so obvious in terms of hardwoods versus conifers, maturity of landscapes and settlements, culture, architecture, and antiquities, that I soon realized the one will never be the other. Yet certain features remained evocative—chiefly, the climate and the colors. That it rains a lot in England and Washington comes as no surprise to anyone; nor that both mean green, green countryside. Rain makes green, as in Ireland and Oregon also. Both British and Northwest landscapes generate a solid and similar green gestalt. So in my travels back and forth I felt comfortable, jet lag and culture shock alike buffered by the soft mental bed of moss and grass that lay at either end.

William Blake described England as "a green and pleasant land." Many authors have since agreed. W. H. Hudson expanded on the idea when, in *Afoot in England*, he wrote of the River Otter in Devon as "the greenest, most luxuriant [place] in its vegetation . . . where a man might spend a month, a year, a lifetime, very agreeably, ceasing not to congratulate himself on the good fortune which first led him into such a garden." Such are my feelings about Gray's River.

Yet, as the Old and New World shades of green run together, I find that they are not entirely compatible. Not so far apart as oil and water, still they have some trouble mixing at the edges. This has to do with fundamental differences of ecology and evolution between places long dominated by humans and those only recently wrested from the grip of nature. In this case, it has much to do with colonial removal of resources on a grand scale and in a manner that dictated many decisions of the future.

Beyond my placid valley stretch thousands of square miles of forestlands stripped of their timber. Almost the whole of the Willapa Hills has been logged off during the past hundred years. The same sort of deforestation occurred in Britain but stretched over a thousand years or more. Of course, much of the forest of Willapa comes back in ranks of even evergreens. Yet the signs of heavy-handed use remain the starkest field marks of this land, in elemental contrast to the gentle husbandry that makes the English scene what it is. The comparison shows how a land may be at once both pleasant and ravaged: it all depends on where you look.

As I look out my bedroom window, I see the signs of both. The green velvet meadows run up against Elk Mountain—long ago logged and now forest again but recently marred by an ugly clearcut that sits like a bad scratch in the middle of the scene. When I settled here, I resolved to seek out both the biology and the beauty that must persevere in spite of such scars.

This is the plan of the book: to describe the Willapa Hills and the wildlife they support, both native and alien; to examine the impact of intensive forestry upon the land and its life; and finally, to assay the ability of organisms (including ourselves) to survive in the aftermath of massive resource extraction. Each major topic—the hills, their denizens, human impact, and survival—receives four essays. Throughout, questions of biogeography, ecology, and evolution in the wet, wintergreen world find their way into the text.

As a disclaimer, I wish to be explicit in saying that no part of this book is to be taken as in any way condoning or abetting the sort of steep-slope, clearcut logging that takes place in large parts of southwestern Washington. While much of what I have to say speaks in sympathetic terms of a community based largely on a logging economy, my sympathies lie with the people and the woods, not with the companies that have used them both with equal disregard. When I write of the beauties and biological interest of the logged-off landscape, it is in spite of the devastation that these things exist, not in any way because of it; and much, much more has been lost. Any attempt to recruit portions of *Wintergreen* in favor of regional timber-stripping practices will be ipso facto misquoting the book and taking the work out of context.

That clear, *Wintergreen* does tell a tale of resilience and a chronicle of toughness. Like an old forest, the towns of these hills are senescing—growing old, losing vigor. Whether senescence of human communities must lead to extinction or may (as in a forest) forecast regrowth, I do not pretend to know. Resourceful people, loving their land, are capable of much. But when a green and pleasant land becomes run-down and ravaged, evolution takes over.

The decisions being made now will determine whether nature takes back Willapa or agrees to share it with us on more equitable terms than we have accepted in the past. Clearly Willapa is a metaphor for wasted lands everywhere. I hope that by framing some pictures of this green and damaged land, these essays will help a little in our efforts to co-evolve with the rest of the living things of the planet.

# Part I
## Rain World

# In the Green Land

One day late last autumn, I abandoned my rural retreat for an afternoon's book work in the nearest city. Winter was closing in as I viewed the first of November from the warm, dry interior of the Longview Public Library. An autumnal confetti of leaves blew across the green, from the oaks and maples beside the art-deco post office to the stoop of the neo-colonial Monticello Hotel. Cold air fell out of a low sky the color of old aluminum, pulled itself up before hitting the ground, then raced along the sidewalk as a bitter wind. Cars squirted along the wet road with yellow leaves plastered to their windshields, while the few walkers on the scene hastened toward indoors. Mellow fall, seemingly spooked by Halloween's howling airs, was clearly in retreat.

In short, a typical northern autumn day, all damp discomfort and chill. No different from November 1 in Helena, Denver, St. Paul, or Buffalo. No comfort to be taken except from the fleeting pleasure of the colored leaves, or the contemplation of closing doors behind you. Melancholy rules, and it's all downhill from here until spring. Or so it seemed that afternoon in Longview.

But winter here, I realized, is different after all. Beneath the shifting carpet of color shed by the trees, the grass glowed green as ever. Nor would it fade to yellow and brown as winter matured, or stiffen into wiry, colorless tufts with the frost. In fact, the frost seldom comes to southwest Washington—or if it does, it tickles instead of torturing the green plants, most years. In fact, winter itself never really comes—at least not a Laramie or a Hartford winter, not even a Wenatchee or Spokane winter such as eastern Washington feels.

When the deciduous trees have all disrobed, the green remains. And the gathering winter wind blows rain, or sea mist, or papermill vapors, or drops away altogether into a still dampness. That chilly first of November could have been the coldest day of the year. Out the window behind me lay the rose garden, whose brightness on this dull day came chiefly from animated bigleaf-maple leaves, blowing through and snagging on the thorns. But a few roses bloomed still, and they would continue to offer a few of their pastel surprises throughout the coming months.

All this is to say that the maritime Northwest differs from other northern places in some fundamental ways. It is a soft green place where rain rules and mildness moderates the proceedings; where the grass grows in January, and the airways and waterways run together in a near-constant interchange of water and mist; where ferns and moss swaddle all surfaces left out in the weather for any length of time; and where the rivers and the sea and the clouds conspire to lend the land a verdancy that never quite runs dry—this is a wintergreen world.

Maps generally orient themselves so that north lies at the top. We are not used to maps that make us look upward at south. Yet I know one such map, a favorite of mine. Published by Tilth, a regional association for sustainable agriculture, its legend says: "On a rare day of partial clearing, clouds separate to reveal the Maritime Northwest." The region, portrayed in green on the map, runs in a gentle crescent against the Pacific from Cape Mendocino, California, to Cape Scott, Vancouver Island; and from the British Columbia Coast Range, southward along the Cascade volcanoes, past Vancouver, Seattle, and Portland, to Mount Shasta.

Thus, the maritime Northwest: "On the east, the Cascade Range protects it from the thirsty plateau. On the west is the Pacific" (says the caption, in blue, engraved on the parting clouds). "Southward the Siskiyous and Trinity Alps palisade the Maritime Northwest against the bare brown hills and burning plains of California. Northward, though maritime climate persists, agriculture ceases, turned back by mountains that rise from the surf."

All this is the rain world. This book, however, deals with only a portion—a mote in the middle of the map of the maritime Northwest. This piece of the region I call Willapa, for the Willapa Hills—the clump of Coast Range mountains lying between the Cascades and the Pacific, the Olympic Mountains and Oregon. Neither the wettest nor the driest part of the rain world, Willapa may be thought of as a biological, climatic, and geologic mean for the region, unusual only in its anonymity and the thoroughness with which it has been exploited. Lacking any national forests, parks, or other reserves, most of the Willapas' wood has long since been carted off, the hills themselves greatly scarred in the process. These essays concern what was left behind. The way to Willapa will show us much about the nature of the Northwest.

I think of it as a land sandwich: soggy on one side, crusty hard on the other, all-veg in the filling. Socked in between the sea and the western mountains, Willapa is defined more by the dramatic features on all sides than by the subtle punctuation of its own hills. Otherwise it resembles the region to which it belongs. The Pacific Northwest shares a common climate and verdure with other moist and mild places such as the British Isles, Tasmania, and New Zealand. The sea, the facing peaks, and how they trade the rain make up the elements of this benign kind of region.

In a sequence familiar by now to every eighth grader, moisture rises off the ocean and flows eastward until, rising against the faces of the cool mountains, it condenses and drops as rain or snow. So the eastern Northwest becomes a desert, the westernmost part of a rain forest, with many ecological gradations in between.

From this familiar model, one would think that all the rain drops at the foot of the Coast Ranges. Actually the clouds distribute their largess all the way from the seacoast to the Cascade Crest, then dribble a little over the summit as well. But the coastal lowlands, the Olympic Mountains, and the vague ranges within fifty miles of the beachhead receive the bulk of precipitation—as weather-station figures show. From 50 to 250 or so inches of rain fall on each square foot of the region annually. At Gray's River, in the southern Willapas, the rain gauge catches around 116 inches annually.

I have heard that Eskimos and Aleuts have many words for snow. Perhaps the Indians of the Northwest coast—the Haidas and Nootkas, Kwakiutls and Bella Bellas, and others—have as many terms for rain. We dwellers of the rain world should have a whole lexicon of precipitation, but we do not. Perhaps we haven't been here long enough, but our vocabulary differs little from other places in this regard. One's neighbors speak of Scotch mists, gully-busters, and "this damn rain," but these terms are not endemic. Where are the autochthonous words—homegrown, born of mixed Indian and immigrant ancestors—that should exist to describe each species and subspecies of raindrop?

There should be words like "mossehurr," to indicate that soft-falling rain that soaks and nourishes the mossy mats without quite wetting the hair; "virkkaplotz," for the mega-drop rains, the falling puddles that set the slugs to slipping off their alder-bark perches and concuss the leaping salmon; "Ratta-tatta-tattarain," for the driven, diagonal, Gatling-gun volleys of pure leaden wetness, rain of the most soaking sort.

Both the immigrant Norwegian fishermen and Northwest Indian whalers knew a kind of sea storm brewed from elements of tradition stirred with reality. The Norse word for waterspouts (*skypumpe*) really should have gotten together with the Kwakiutl name for their mythic and totemic ocean-going serpent, the Sisiutl. "Sisiutlskypumpe" could apply equally to the fearsome waterspouts that seem spat out by sea beasts, and to the hungry fogs they spawn, that run up river mouths to devour the land with a foaming, salivating sort of rain. We could do with a set of names like these for the storms at sea, the riverine damps, the various vehicles of falling water. But we have no such words.

In the absence of our own onomatopoeic, hybrid monikers for moisture, we might at least have borrowed from the British. Shouldn't such a perennially damp people have evolved a taxonomy for rain? Perhaps they have, but in my experience the best they can do for a North Sea Sisiutlskypumpe is to call it "a bloody filthy night out there."

So far, I have learned only one local word relating to rain: freshet. Not purely local, of course, it appears in *Webster's* as "a flood or overflowing of a river, on account of heavy rains or melted snow; an inundation." The *Oxford English Dictionary* concurs and adds that a freshet may be "a stream or rush of fresh water flowing into the sea." Naively, I thought that a Gray's River freshet was the same as an OED freshet—that any rain big enough to make the river overrun its banks deserved the term. But it's just not so. I got the final word on the best of authority.

When I need enlightenment on matters of local history, natural and otherwise, I make for the Gray's River Post Office. Jean Calhoun, postmaster and half-century resident of Gray's River, usually has the answer for me. So one moderately flooded day, I asked her, "Jean—is this a freshet?"

"Heck, no!" she corrected me. "It's too cold, and the wind is from the southeast. A freshet comes on a warm west wind, and looks like this." She shoved a recent telephone book under the window grate. On the back, the Gray's River freshet of January 4-6, 1914 was depicted. The then more substantial town really had its feet wet, and its belly and its chin, the whitewashed church and the Ferndale Creamery looking as if they were afloat. (Carlton E. Appelo, general manager of the Western Wahkiakum County Telephone Company, also serves as local archivist and historian. In a creative move to preserve local lore and serve it up to his captive audience, Appelo devotes part of each phone book to recording the area's colorful history. These double-duty documents stand by to illuminate points such as the one in question.)

So I was educated about freshets—not in the OED but in the Gray's River Post Office—big, long rains on a warm west wind that really raise the rivers and bring superhigh tides, the two together deeply flooding everything in reach for a few days.

Just as we need a vocabulary for rains, we should have a nomenclature for rainworld landscapes. "Temperate Rain Forest" is the best we can do, for the humid, mostly coniferous woodland that is the natural climax and secondary growth of the extreme maritime Northwest. The famous rain forests line the lower valleys of the rivers running out of the west side of the Olympic Peninsula—the Queets, Quinalt, Bogachiel, and, best known, the Hoh. All these were saved as virgin rain forests by slender fingers of Olympic National Park. Franklin D. Roosevelt's fiat, Congress's laws, and Justice William O. Douglas's publicized protest hikes all helped protect these great foggy forests from the saw, while those all around them went. Even today, fervent conservationists must struggle against ploys to raid the rainforest corridors.

Though far from virgin, Willapa is rain forest also. Secondary or tertiary rain forest, it is true, lacking the depth, size, and integrity of a Queets or a Hoh. Still, creases and wrinkles in the abused skin of the Willapa Hills show the field marks of rain forest: garlands of *Usnea longissima*, pale green old-man's beard lichen, hanging from hemlocks like Spanish moss; tresses of chartreuse club moss padding the boughs of bigleaf maples, swinging from the limbs in the ratta-tatta-tattarain and luxuriating in the mossehurr. Sprays of licorice ferns sprouting from the collarbones and patellas of the lumpy maples, and vast banana slugs patrolling the fecund forest floor or negotiating the nurse-log greenways. By these terms, much of the maritime Northwest qualifies as rain forest with greater or lesser precision.

What about rain valleys, rain meadows, rain ridges, and rain rocks? All of these need sobriquets too. Perhaps my sense of "need" here comes from having spent a long time in England, where all the features of the ancient landscape have their own names—bestowed in antiquity, preserved by Gilbert White, Henry Williamson, and Edward Thomas, or invented more recently by J. R. R. Tolkien or Richard Adams. My own surname, Pyle, derives from "pil"—a West Country appellation for a small tidal stream (or freshet?). I suppose such terms arise from centuries of closeness to nature; perhaps we have had neither the time nor that necessary nearness to have evolved many of our own.

Among my favorite British landscape terms are those for winter-flooding meadows. Like T. S. Eliot's cats, they have three names. "Holme" is the Old Norse term; "water meadow," the name applied by graziers and hayers. And more recently, a campaign to conserve these special, ephemeral wetlands called itself "Save Our Silver Meadows." Whether "silver meadows" is an old name or was newly coined for the campaign, I do not know—but I think it is a particularly lovely name for an equally beautiful feature of the countryside.

My windows overlook silver meadows for several months of the year (jade or golden for the rest), but here they have no such name. We just call them hay fields or pastures, with or without ponds, flooded or high and dry, green or yellow. Our taxonomy of rainscapes is indeed impoverished, not at all adequate for the topographic richness of the rain world.

But if our vocabulary is wanting for weather and landscapes, how much more so for color! Particularly green, the one color for which we should be richly endowed with descriptors. As Irving Petite has written, in western Washington "one is *inside* the jungle, and often seems to be going down for the third time, drowning in green." Yet we remain impoverished for ways to say it. Peter Steinhart, writing on color in *Audubon*, points out that "Nature gives us no bright-green dyes, and that could account for the paucity of common names for greens."

Others have noticed this lack. Northwest poet Melinda Mueller was once a guest of my house sitters while I was abroad. On my study wall, she noticed a map of the world's biogeographic provinces. It inspired a poem, later published in the Winter 1984/85 issue of *Fine Madness*, in which the poet describes

*... the rainforest. Green,*
*green, green,*
*green ... no two greens alike, but there are*
*only so many adverbs (a yellow-green*
*leaf floats just under the river's gray-*
*green surface). What we lack*
*for the rainforest is*
*twenty words for snow,*
*but not for snow.*

(It seems the Eskimos' descriptive vocabulary is rather well known.)

I suppose if we were to peruse color-theory manuals, housepaint-chip displays, or racks of artists' tints, we could come up with a long list of grays and greens. But few of these are in common usage in a land painted from a palette of grays and greens. A fern is a fern is a fern, and perhaps we suffer too much from sensory fatigue to fully appreciate the array. I am surprised that we of the rain world don't all see red when we close our eyes, as I do when I look away from these green words on my computer screen.

Peter Steinhart has another thought on the matter, one I like: "Color is a pleasure that is private and idiosyncratic, and when we gush over a scene we tend not to trust our descriptions of colors. It is not because we fail to see colors clearly. It is because color hasn't yet arrived in that part of our minds which honors words. And it is because our sense of color has crept beyond the realm of nature."

Perhaps because there is no apparent utility to richness of language out in the country, far from the salon, the studio, the coffeehouse, or the classroom, simplicity of expression has taken over. This applies not only to grays and greens and meadows and mists, but to all aspects of daily life. People speak directly here. And, while one may mourn the absence of honor for words, simplicity and directness have their merits. In the rural rain world, intellectual pretension will get you nowhere. It's as if all the rain has washed away the extra words and gestures that clutter up drier and more sophisticated places such as Portland and Seattle. So it is not surprising that few words exist for the green land's damp and hues.

As in the map on my wall that stirred the visiting poet, biogeographers use a range of colors to imply the patterns of plant and animal distribution. Biogeography suggests ways in which living things array themselves on the earth's surface, or would in the absence of human disturbance. Principles govern biogeography, laws constructed from evolutionary possibility and ecological need. One of these principles says that organisms become less common and more specialized near the edges of their ranges. Thus, the European swallowtail butterfly (*Papilio machaon* L.) occurs abundantly in many kinds of habitats on the continent, yet only rarely and in a few remnant fens in Great Britain. Having been thoroughly glaciated quite recently and then isolated between the North and Irish seas, the British Isles are pretty marginal overall in terms of biogeography. Many British animals and plants occur on the edge of their tolerable range, and it has been quipped that this may be true of humans as well.

Hardly, in fact. But the joke (probably spoken by a Frenchman or a cold American) has an edge to it: a cool, damp climate may lack the extremes of a truly marginal place such as Greenland (where only six species of butterflies live, in contrast to Great Britain's sixty, North America's six hundred plus, and Peru's six thousand); yet it imposes on people and other animals the need to specialize. and it excludes many sun- and warmth-loving types. Hence the swallowtail's liking for the fens, where it finds conditions most favorable for survival; and Britons' well-known penchant for woolies and tea.

If Britain is a bit edgy, how much more marginal is this rain world of Willapa—with three or four times the rain, yet without the support systems provided by an ancient, mature culture? The surrounds of this moist orifice of the West, the mouth of the Columbia River, have been eschewed by many and deserted by many others. Whether repelled by the rain and the gray, put off by provinciality, or forced to leave by the bust that followed the early resource boom, would-be greenlanders stay away in great numbers.

So too with biologists. With the rich, protected Olympic Mountains to the north, the extensive, wilderness-and-park-studded Cascades to the east, and the more impressive Oregon Coast Range to the south, why should anyone come to the little, stripped Willapas in search of wildlife or natural history?

True, fisheries biologists abound on the Lower Columbia and its tributaries, and shellfish researchers along the coast and estuaries. One or two naturalists have appreciated Wahkiakum and Pacific counties' bounty of salamanders, a botanist prowls the cutover hills for rarities, a handful of birders pass through and sometimes stay. Hunters and managers pursue the elk and the deer, fishermen seek the steelhead, salmon, and sturgeon. But on the whole, nature lovers ignore this seemingly dull and impoverished landscape. Seekers after wild rivers, forest glades, and alpine vistas go elsewhere, never even thinking of the Willapas. You will not find a guide to "101 Hikes in the Willapa Hills" in any bookshop.

Nor have these hills inspired much in the way of fine arts. I cannot speak for the other side of the Gray's River Divide, but I suspect the overall artistic output of the Willapas has been much the same from one end to the other: modest. Good local histories appear, in telephone books and elsewhere, and a scant handful of prose and verse writers have scribbled here. Yet there is scarcely a literature of this land. The same cabin fever, perhaps, that spawned the winter ceremonies of the Chinook Indians gives rise to some surprising music and theater. A couple of potters, one or two photographers, painters, printmakers, and not many others have drawn inspiration hereabouts. The rain world of Willapa has yet to produce its Hardy or Kesey, Constable or Graves—and it probably never will. The great goal was cutting down the big trees. After that, modest hopes and diminished expectations seemed to take over.

The rural rain world seems to be a place whose promise flushed and faded early, where people live who were stranded by history or came by accident and got stuck. But that doesn't quite tell the story. Thousands of Scandinavians and other immigrants came here of their own will and intention, for its openness, promise, and resemblance to home. And for certain others, the wet Willapas seem to have a genuine attraction all their own.

I am one of these. I find the simplicity disarming. Many members of the community are openhearted and open-minded, perhaps because in fighting for their survival they are willing to welcome anyone and listen to all ideas. The isolation has its own rewards, among them fresh air and a plenitude of peace. Ready antidotes to parochialism come in like-minded friends, books, visitors, and trips to town. However, I find that I take to the city less and less as the local fit becomes more and more comfortable.

As a biologist, I have been surprised by nature's rewards in these hills, though they never come without reminders of abuses of the past and present. The ability to walk out of my door and have woods and wildlife immediately around me has become terribly important. And as a writer, I find the pervading grayness to be salutary: a gray day in Gray's River provides few distractions. Annie Dillard is of course correct when she says that serious writers must take a broadax to their lives; that's harder to do in the rich, titillating city. These attractions explain my voluntary rustication, after an urban upbringing.

As for a midcountry mountain lad landing in the maritime Northwest, I have to say I came for the rain—and the green. As I grew up in dry Colorado, my Seattle-bred mother told tales of the moist Northwest. My grandmother and great aunt, pioneer teachers in Washington, reinforced visions of a viscid, verdant place, dripping with mosses and ferns. When they read me stories of the Emerald City of Oz, I equated it with Washington State. Day trips into the Rockies found me seeking out the few ferns and mosses that grew in shady spots and imagining acres of them on the fabled Olympic Peninsula. At last a visit in

1964 with my mother, to see the University of Washington, brought me directly into the rain forests of Mount Rainier and Hood Canal. I was entirely entranced.

One more brown Denver March made up my mind: I headed for Seattle in the autumn of 1965 as an undergraduate. I took up residence near Ravenna Park, the same forested ravine that my mother grew up beside. Whenever possible, I used campus paths grown round with ivy and sword ferns. I always believed I had a gene from my mother that conveyed her passion for fern-strewn ravines and rotting mossy logs. With only a shade of Colorado longing, I became an unrepentant denizen of the rain world.

Field studies and friends later introduced me to the Willapa Hills. They struck me, somehow, as both sinister and inviting—at any rate, compelling. The fact that virtually no butterfly sampling had ever been conducted was attractive (before I realized that there were precious few butterflies to be found). In fact, little of anything had been done here, it seemed, in the way of biological investigation. It was a virgin land in this one regard, if in no other. So I suppose I was attracted by the same ethic and sense of possibility that brought the Finns and Swedes who replaced the Chinooks and changed the place so—to exploit it, in my own unorthodox and latter-day way.

Beyond that, at last I had enough green and enough rain to wash out all the Denver dust and boyhood browns. And acres and acres of ferns and mosses. And all the slugs and snails I had dreamt of finding as a landlocked boy conchologist in Colorado.

But I also found that I had been misled: the rain is not omnipresent. The sun also shines. And sometimes it snows. Over Yuletide, Boreas presented his credentials at the door and one morning I awoke to a pallid valley white with snow. It did not last. The cold stole a shade or two, but the green remained. I figured our annual allotment was past. Then the next time I went to the library in Longview, same chair, same window, the snow really flew and stuck. Not at all green, but black and white, the scene resembled one from a movie gone to old stock for effect.

Smugness about prevailing conditions begets uncomfortable exceptions. Yet, I reckoned, what is novelty to some, messy for others, a driving hazard, or a pretty pleasure, and all the above to me that day, is annual reality to many. Unless the Pleistocene is biting back for having been sunk behind the Recent for the present (as nonsensical as the term "postmodernism": how can something be after modern, or currently recent?), I figured we would see all this Arctic weather pass in a week or less.

And we did. The sun came out for a solid month. High pressure blew in from the sea, the river ran like a quicksilver snake to the bay, and frosty nights turned the valley as brown as it ever gets. All the while I wrote in my shirt sleeves on the porch, and the cat and the cattle basked in the sunshine. I wondered which of us appreciated it most. Of course, we were all fools together in a fool's paradise, the cat, the cows, and I. I knew, and they probably suspected, that we were likely to lose the sun any day. The rains would reassert themselves. And we would watch the rest of winter drip soggy into stretched-out spring, called on a carpet of deeping green.

And so it went. For in the green land of Willapa, the long-range forecast is always the same.

# The Hills of Willapa

I suppose it is a hallmark of all but the most charmed youths that they are spent as much longing for what is not available as enjoying what's on hand. So it was that in my boyhood I longed not only for the rain forests of the distant Northwest, but also for the much nearer Rocky Mountains. For though I lived in Colorado, I grew up on the wrong side of town—from the mountains.

My good friend lived in a west Denver suburb, abutting the foothills of the Rockies. He could walk out of his house and a short stroll would take him to Green Mountain—which, I realize now, wasn't very green in absolute terms—to hunt the green hairstreak butterflies (which were).

By contrast, my neighborhood hinterlands were much more plain: weedy old farmland and ditch-banks on the backside of Denver's hem. I always thought of the rich, extravagant mountains as the place to be. Family drives and fishing trips would take me there but too infrequently, and I always imagined someday living in one of the mountain canyons where the swallowtails flew in June.

Now I live in a mountain range. And in a way, the joke is on me. For these are no mountains in the mold of the Colorado Rockies, spilling with grandeur and color. These hills are as subtle, in their way, as the high plains vestiges where I stood as a boy gazing longingly across Denver's gathering haze at the pale forms of Long's and Pike's peaks, Mounts Evans and Audubon.

Not that the mountains of the Pacific Northwest lack extravagance. I could have had my green and my spectacular mountains both, had I settled in the Cascades to the east or the Olympics to the north. But circumstance brought me here to the Willapa Hills—the gentlest adjunct of the western ranges.

When one thinks about mountains, one thinks of significance. Mountains are *about* significance. Mountains sign the power of the ages, to move and shift the very stuff of the earth. They signal the life of the land, since the stones are the bones of the beast that rolls in the rapture of orogeny. And mountains signify relief from the flatlands, to anyone who as much as sights their heat-hazed forms across the monotonous miles of the plain. But as mountains go, the Pacific Coast Ranges are short on significance. And of the Coast Ranges, the Willapa Hills of southwest Washington are the most insignificant of all.

Not far away one of the outliers of the range projects above the floodplain of the Naselle River. Due to its strategic position on the southwest shoulder of the Willapas and its somewhat unusual rocky top, this palisade has long been used as a pedestal for electronic communications devices and is known as Radar Hill. Most of the signal buildings stand empty now, but the outlook remains the best around.

Since an Air Force setup on top was abandoned, the reasonably good dirt-and-rock road to the top has been open to anyone who wished to ascend Radar Hill for the view. I often drive up on sunny days to seek hill-topping butterflies and, in winter, hibernating butterflies in the many old concrete structures left behind—perfect for the purpose, full of dark spaces accessible to insects via holes in cinder-block walls pecked out by the points of the ice-pick wind. The incessance of that wind usually defeats my purpose. But it may also clear the air, so that if the horizon's haze of slash-burn or wood-stove smoke, volcano and nuke steam, or sea mist doesn't block it, I have a fine prospect of the Willapa Hills and all the landforms that hold them in.

Here is what I see on such a day: to the north, in the distance, the ghostly white jawline of the Olympics draws the east-west edge of sight. The Olympics are significant mountains, looking like the Grand Tetons shorn of half their vertical exaggeration and stretched out twice their length. On the east reclines a blue-green backbone studded with great white vertebrae at intervals. This is the Cascade Range, with the Cascade Volcanoes protruding above the rest—Mount Rainier, Mount Adams, what's left of Mount St. Helens, and Mount Hood in Oregon. Following the line of sight around to the south, still in Oregon (which is very near at this point), I see the ruffled shapes of the Oregon Coast Range. Non-Tetonic, it nonetheless puts forth the impressive shapes of a real skyline. Saddle Mountain, a great three-humped camel, and sharp-tipped Onion Peak out toward the coast look like real mountains. Then I turn to view the west, and I see the ocean—first the oystery indentation known as Willapa Bay, then the thin green nail of the Long Beach Peninsula, finally the Pacific itself, its breakers a lumpy white line beneath the sky.

These, then, are the several horizons bounding Willapa: the Olympic Mountains, the Cascade Mountains, the Oregon Coast Range, and the Pacific Ocean. In fact, the natural region of the Willapa Hills has nearer, more precise borders. When you look out from Radar Hill, you see the land drop away into troughs before it rises again to the distant ranges of mountains, hills, and whitecaps. These are the riverine and glacial valleys between the ranges: the natural, biogeographical boundaries between earth districts on a fine scale.

The vague northern trough corresponds to the long valley of the Chehalis River. The flatlands hinted by the haze between our lookout and the snowcones of the Cascades to the east are the floor of the Puget Trough. This glacial tongue-groove slips on down to become the Willamette Valley in Oregon. But before it does, the Columbia River bisects it and shoots out to its release from dams and canyons at the seashore. The tidal Columbia forms the southern boundary of Willapa, with the Oregon Coast Range rising on the other side. And on the west, the Willapa Hills drop into the bay that shares their name.

Here, then, we have a region of about fifty miles on a side, shaped like some great bracket fungus conk that scallops the ridges and furrows of its supportive bark. It differs from the foothills of the Olympics and Cascades not so much in kind or aspect as in the fact that those *are* foothills: they rise to something greater than themselves and therefore share in the substance (geological, scenic, spiritual) of high mountains. The Willapa Hills rise to nothing more significant than themselves. To appreciate them requires that you recognize this fact and adjust your expectations accordingly. What you see is what you get.

We are still up on top of Radar Hill. We have found in the near distance the edges of the range in question and determined its extent. Now look more closely: what do you see? Hills. A sea of hills of many altitudes and attitudes, generally ranging from little to medium—a few hundred feet elevation to two or three thousand, tops. A handful of the "peaks" jut more prominently than others, and these usually became the fire lookouts: Hull Creek Lookout, K. O. Lookout, Ten, Walville Peak, and, highest in the range, 3,114-foot Baw Faw Peak. Two prominent ridges stride across the eastern frame, both called Gray's River Divide. Myriad creeks and streams and a few rivers cut the range up into valleys and ridges, then run around to join one another and form the river-bounds just named.

More recently if not more literally, saws of loggers cut up the hills as well. In fact the first thing most people would notice from Radar Hill, before the sea or the volcanoes would be the fact that these hills have been logged all to hell—positively butchered. And they might then look no farther, but drive back down, shaking their heads in disgust, and head off to the parks and forests of the higher ranges. I am convinced that most naturalists who ever take a first look do just that and don't stay for a second or a third. My inclination at first sight was the same. But out of sheer propinquity, I took to these hills again and again. My distaste for some of the logging practices to be found here never diminished, but my ability to edit them and see between the lines of the clearcuts grew.

Those who abandon the lookout retching at the flayed forests without looking closer may miss a great deal, without even knowing it. One of the only direct benefits of clearcutting, for example (to anyone other than the profit makers), is the fact that it lays the land bare for geographical appraisal. I remember several geology professors at the University of Washington repeating that, although the Pacific Northwest climate gave them plenty to gripe about, its logging industry furnished them with rocks and mountains nearly freed from their cryptic forest wrappers. So that while their fieldwork might find them with cold rain running down their collars, at least they could see the landforms more clearly than their pre-logging pioneer colleagues might have been able to do. As we will see below, earlier geologists found their conclusions altered by later workers. This in itself is not surprising. But to what degree it was due to progressive deforestation versus better fieldwork, we cannot be sure. So accept with me for a moment that, without being complacent, one can edit the scars on the skin of the land long enough to study the bone structure beneath; and later I will show you how to find hidden beauty marks as well.

Of course, in reality, the rapid regrowth of scrub, brush, and trees in the well-irrigated Willapas means that the actual soil and bedrock seldom show naturally. Where they show more frequently is in roadcuts and where heavy equipment and steep-slope logging leave bare patches open to erosion. Here the soil (on its way into the salmon-spawning beds of streams and out to sea) lies exposed to vision in many places. The relatively few rock outcrops of the region grow trees slowly, so once logged they remain open to view—like Radar Hill itself. And the even-aged stands of managed second-, third-, and fourth-growth timber (although ecologically impoverished and vulnerable) do reveal the underlying shape of their substrate just as his well-sculpted flattop defines the cranial crest of the high-school coach.

So we can see these hills. What is their nature? Three sources, differing from one another as they approach the present, help to illuminate the picture. *Roadside Geology of Washington*, by D. D. Alt and D. W. Hyndman, *Cascadia*, by Bates McKee, and a 1926 geologic map of southwest Washington by C. B. Weaver all speak to the underlying structure of the Willapa Hills. The latter map, largely for its pleasing, aged tones of salmon, tan, beige, and olive, hangs on my study wall. I like it and would be happy to relay its image to you as the true word, were it not for the many ways in which it differs from the later sources. Anyway, I am fond of them as well: the late Professor McKee's *Cascadia* conveys his rich enthusiasm for the topic, which I remember so well from his classroom. And the *Roadside Geology* endears the user with the authors' clear, interpretive style and its easy-to-grasp diagrams of difficult features. I am struck by the degree to which the three sources differ in detail. But we need not be concerned with these arcana here. In essence, they agree on the basic configuration of these hills, so simple by comparison with the Cascade or Olympic mountains.

To speak in simplified terms, the Willapa Hills consist of a layer cake of igneous basalts and sediments, moderately folded and interbedded. The main body of the cake consists of undersea-erupted basalts that were a part of the old oceanic crust. The frosting is mudstone and siltstone laid down on the crust when it lay below sea level. Columbia River basalt flows protruded on the south and east some fifteen million years ago, complicating the layers. Riverine, lacustrine, and beach deposits marble the pattern with younger sediments, often poorly or not at all lithified. Riding on the same kind of ocean-trench debris that forms the Olympic Mountains, the Willapa basalts and sediments lie well above sea level and far above the current level of ocean crust. The hills represent ancient uplift followed by the lapping of rivers at the frosting and the peckish licking of the eons at the cake itself. The result, a standard landscape for a region rich in rivers and vegetation, is called ridge and ravine topography.

Up north, a big surprise called the Olympic Mountains popped out of the cake. This uprising shoved up a series of foothills that look a lot like the Willapas, and in fact run into them between Gray's Harbor and Puget Sound. To the east, the cake bumped into the main course of the Cascades, whose recent emissions hide the border for the most part. On the south, another portion of crusty cake continues down through the Oregon Coast Range. So in actuality the Willapa Hills may be said to extend as a geologic unit from the ocean to the Cascades and from the Olympics to Oregon—just as it appears from Radar Hill.

Yet, as I have said, it is more convenient to consider this geographic province bounded by the valleys that cut through the mists and into the cake, making a smaller portion. We cannot see the underlying bedrock that defines the ultimate limits and alliances, nor do we react with it. In any case, by geographic criteria, the Willapas simply serve as a lumpy bonding agent between better-defined provinces on three sides. What we can see is a limited chunk of land, full of ridges and ravines, contained within a satisfying set of lowland landscape lines. The cake, picked at and nibbled beyond recognition, is still a solid piece of stuff lying within the rims of its cake plate.

The ravines, laced as they are with logging roads, allow for many alternative routes into and across the Willapa Hills. You can view the hills from a lookout, but to sense their true nature you need to drop down into the ravines, climb them to their ridges, then drop and climb again. Owing to the nature of the ocean crust as modified for log trucks, this may only be done deliberately. So, applying equal care toward not getting stuck in a remote place and toward genial observation, one comes down from the hills enriched as to their flavors and facets.

Now seven years in the Willapa Hills off and on, I have made many crossings and explorations. My journals contain scores of pages reflecting sights and smells, images and shadows of the hills. Some of these notes will find their way into these essays. But in attempting to describe the Willapa Hills with some immediacy, I felt the need for a fresh penetration. So on the last day of 1984, I chose a certain passage over a portion of the province, river to river, ravine to ravine, ridge to lookout, and down.

One rare sunny afternoon in midwinter, I wanted to see its sun set on the year from a vantage deep within the Hills. So I left the highway via Salmon Creek and began my car climb up into the Willapas. A forest of spruce and cedar lined the road, held back in the shadows by a frosty verge. Pale green moss hung in cobweb strands from the outstretched Sitkaspruce boughs. Backlit, they resembled so many slender arms dressed in tattered luna-moth silks, supplicating the season; and ardently, it appeared, for they dripped from the armpits. Where the sun escaped their grip, it fell on sparkling dewberry and moss-pile flooring.

An avenue of frosty spruce and hemlock, Christmas trees flocked for real, ran down the 5000 Line to the Salmon Creek Road. Crispy white salmonberry bushes stood above overbent sedges, their edges describing white dual carriage-ways, their fruited spikes sticking up sharply. Across the road, horsetails lay bent and busted beneath frost and age. I came to a slash-burned logsite, never a surprise in these hills. An old stump had been burned out from beneath a scorched hemlock root system, whose tree in turn had been logged, leaving a black octopus walking on tiptoe across a halftone moonscape.

Leafless, red-budded, white-crusted alders formed a fore-stage chorus for the conifers: a graceful curve of slender trunks and intermingled tops, all the same height, hiding the feet and middles of the conifers from view. In fact, the hemlocks eventually will expel the alders through natural succession and shading. The foresters would prefer not to have to wait for that to happen, so ordinarily they spray the alders away with herbicides after clearing the conifers. Alders, usually considered weed trees by foresters interested in evergreens only, nevertheless lend much of the beauty (not to mention nitrogen) to the hills. A lower story of yellow withies—salmonberry canes—reached out to the road, punctuated by scarlet wands of osier dogwood. Native red huckleberry shoots lined up with alien, green Scots broom in a Yulish display of bright herbal hues. From the valleys below, the hills look quite humpy, if not ordinarily pointy. When you get up into them, the relief evens out into evergreen undulations—the gently rolling ridge and ravine landscape of the guidebooks.

Up Spur 5910 to the Fall Creek headwaters. Palpable sunbeams fumbled through a hemlock curtain, turning on the footlights of the frosted road gravel. Steam rose from the roadbed, and salal and sword fern slid through the field layer on a slippery green skein of sunshine.

Now came a spot near the merging of a small, unnamed creek into Fall Creek that is the very essence of Willapa. The stream ran shallow and (for the time being) clear in its ravine-bottom bed. In spite of clearcuts upstream and down, at this spot it could only be called pretty. Great downed alders in rainforest garb lay across it, frosty-mossy bridges to foxgloved islands in the middle or to the salmonberry swales of the other side. Standing alders and bigleaf maples wore moss so thick on their boughs that they brought to mind Eddie Bauer models in down garments, or the Michelin Man crossed with the Jolly Green Giant. Licorice ferns, which curl when dry or very cold but seldom freeze, described pleasing parabolas as they launched from these moss-bound boughs and trunks in clusters. Theirs was the freshest green on the verdant scene.

A woodcutter passed, his pickup loaded with alder cordwood left behind by the big boys. Another alder lover—the only other person in all the woods that day.

On the hills above the mumbling river flat, tall timber somehow still stood. All of the local coniferous species appeared to be present except the rarities such as Pacific yew and true firs. From the slope above the road, an enormous old-growth western red cedar launched its 150 feet or so at thirty-five degrees toward the stream. I daresay it would reach, if it fell, or perhaps make another mossy troll-bridge. Five feet, six, through the base? In its forked top, and among the alder cones and all the surrounding foliage, a mixed feeding flock of winter birds flitted and chittered. Chestnut-backed chickadees and golden-crowned kinglets juxtaposed their near-nothingness beside the immensity of the slanted cedar. Animals of inbetween bulk—Roosevelt elk—did not appear on that occasion. But they might well have, as any passage across the hills runs the welcome risk of encounter with elk, save in elk-hunting season.

White quartz crystals protruded from igneous roadstones, as ice crystals jutted several centimeters from the soil, and hoarfrost from the serrated edges of alder leaves. Yet—wintry as it was—the greens still had it over the browns and whites. The "dead of winter" isn't very dead here. "The greening of America" may be largely forgotten, although global optimists still speak of the greening of the whole world. In any case, greening of the Willapa Hills would be highly redundant.

A few yards farther downstream, however, the rest of the Willapa story came out: above a striking series of riffles and waterfalls, a clearcut tumbled down to the riverflat. The Forest Practice Act allows harvest of merchantable timber right down to the streamside. Here, alders happened to grow along the bank and had been spared. So if one regarded only the immediate shore, the scene still looked attractive. But not beyond. The wholeness of the setting just upstream was gone.

When will it happen at the site of the great cedar? No spot in these hills seems immune from periodic massive disturbance. Understanding that vulnerability is a key to appreciating the essential nature of the Willapas. We can cross from ridge to ridge over ravines and find pretty places undreamed of by highway passers-through. But we mustn't expect them to stay that way. That can be hard to take, and it presents more than a little challenge to the organisms involved and those who would appreciate them.

Heading up Hull Creek, breaking out of another rich and beautiful bottomland, I emerged into expansive clearcuts on steep slopes reaching right down to and crossing the turbid, brushed-up little stream: typical. And then a surprise (small surprises are also typical, and we are thankful for them): an extensive sedge marsh at valley's head. Is this an old, natural feature, or a product of recent cat-damming? Such questions often arise and suggest return trips in summer to look for specialized plants and animals that would indicate remnant natural features. This could be one of the rare, rewarding butterfly habitats I am always seeking.

This big cut would send most seekers back downhill. But, persevering, I came up into the sun with the hills all laid out to Saddle Mountain across the Columbia to the south. And then the old Hull Creek Lookout road appeared. Where it takes off from the logging road, an emerald rivulet ran beside a raft of land solid with sword fern, prowed by a fox-red cedar stump, passengered by a crowd of hemlocks; and off to the side, up the tributary, one grand, gray hemlock stood. A mossy log jutted across the brook before it, giving passage to squirrels and salamanders.

Such a wealth of ferns! If ferns were greenbacks, all the unemployed loggers of the lowlands would have been up there with me. But except for the occasional fern gatherer who serves the florist's trade, ferns' only currency exists in their beauty: nonnegotiable and nonrefundable, lacking even the paltry value of the aluminum trail left by the hunters along the roadside every fall. Yet I am enriched by the ferns in a way that pleases my eyes if not my banker.

Up a ridge road, the sun came through and transformed hemlocks into instant redwoods. And then I saw a flying deer. Charging out of the forest on my right, the black-tail doe leapt from the top of the salal bank and sailed across the road, eight feet up in midair. Four points down and bounding, she descended the ferny slope to my left. Black-tailed deer usually seem a little ordinary to me, though I am always glad to see them, like any wildlife. But this aerial ungulate erased any previous image of the black-tailed deer as a humdrum animal. My heart still racing from the sight, I encountered another doe at the next junction.

As she crossed the road gingerly, with all four hooves on the ground, her coat shone deep chestnut, a rich winter hue of thick pelage.

I got just a short distance onto the old lookout road before snow, mud, and a downed spruce put pay to further driving. That was all right: I wanted to walk ultimately. So I set out near sunset, on foot. Unlike the rocky logging roads made of broken basalt blocks, this abandoned right-of-way seemed like a country lane with a flat surface of needles and dirt, the incline gentle and the snow never deeper than my boots, so the walk was pleasant. There are few trails as such in the Willapas; the choice is usually between one of the great warren of logging roads or obscure game trails through the underbrush. So such a track as this was to be savored.

Halfway up I reached a ridgetop pointed north and looked westerly toward the sunset. A streak like a fresh salmon steak lay along the horizon, as rivers of mist rode up the valleys and stuffed the ravines. The Long Beach Peninsula and the headlands at the mouth of the Columbia glowered, dark animal shapes against the paler invisibility of the sea. Between me and them, hordes of hills tumbled about quietly, as if waiting patiently for the day when the rocks beneath will ride up to make a new mountain range, as they already have in the Olympics to the north.

As I continued, half a moon and the whole of Venus at its greatest elongation came out, side by side. Icicles hanging from moss and roots above the path glittered in their light. I followed my moonshadow on the snow all the way up. A constant stream of animal tracks preceded mine—deer, elk, coyote, perhaps bobcat, and many lesser creatures. We all found the old track easier going than the tangled forest to the side.

No sign of the old lookout tower greeted me on top of Hull Creek Mountain. The small, flat clearing gave view to Saddle Mountain fading fast in the middle distance and to the cluster of lights that was Astoria to the southwest. Bulbs in ones and twos glimmered through holes in the mist in nearer valleys. The dark made it hard to assess the tracks on the apparent playground around me. Young hemlocks blocked any dim view I might have had of the Olympics, and the volcanoes were just too far off to be made out in the moonlight. But easterly and close by, the snow-striated faces of K. O. Peak and its cohorts on the Gray's River Divide stood out prominently. These anticlines push the crustal basalt into the most dramatic peaks we get. And in the dim, I could imagine their striations to be geological, like those sedimentary lines of Colorado's famous Maroon Bells, instead of logging scars.

So I finish this perambulation of the Willapa Hills where it began, up on a lookout. But K. O. Peak is not the Maroon Bells. No ten-thousand-foot wilderness summit, but a logged-off hill barely over two thousand feet in elevation. And in making such a comparison, I commit the same mistake that others make when addressing the Willapas: we try to make them into something other than they are, forcing them to come up wanting. These are understated hills, not very high, made of lavas and mud, and nothing more. The fact that they once supported one of the greatest forests on earth is beside the point since that forest isn't there anymore—it's gone to sunken ships, second-hand furniture, derelict buildings, and yellowed newsprint. These are devastated hills, doing their best to recover, to grow green things in time for the next devastation. A ravaged land, awaiting the next ravages. It is no wilderness; and yet it is wild and elusive.

In the 1930s, a biologist named James Macnab conducted a remarkably detailed study of the plants and animals of a site in the Oregon Coast Range, prior to the first logging in the area. He called his study (not published until 1958) an "Aspection"—as it attempted to characterize the changes in aspect of the forest as the seasons turned. Rereading the paper recently, I was reminded of two things about the more northerly Willapa Hills.

First, that the aspect of the land does indeed change from winter into spring, from the austral to the autumnal. The way I have described the hills might differ quite a lot from an account of the same passage rendered in May. And yet the hemlocks and the sword ferns and the salal, the clothing of the essential green folds of the land, remain much the same. The hills have their seasons, of course; but as much as any temperate place, the wintergreen land of Willapa maintains its essential aspect from solstice to solstice.

Second, that such a study as Macnab's has never been performed in these hills, and probably never will be. Ecologists don't tend to flock to the stubble left by the logging companies, like starlings to a leveled cornfield. On the whole, we like a richer diet. As much as anything, that is the reason for these essays. No "aspection" or ecological monograph, they convey impressions more than statistics. In attending to these neglected hills I try to appreciate them for what they still are, without holding against them what they once were.

Still, in middle age, we look too often to the other side of the fence. Now I live in the mountains, and my friend still lives on the west side of Denver, and he still has more butterflies nearby than I. I suppose I could have moved to Peru. But truth to tell, this is the other side of the fence, and there is no place greener. My grandfather used to have a saying that he repeated at every Sunday dinner: "This is a mighty fine dinner, what there is of it; and there's plenty of it, such as it is." For years I found his Kentucky aphorism cryptic: what's that supposed to mean? Now I understand it better.

The Willapa Hills are like that: fine what there is of 'em, plenty of 'em such as they are. One says it wryly, and might wish they were more, or otherwise. And yet they offer surprisingly much to the observant naturalist whose expectations are not too high, who can edit ugliness, and who is open to the serendipity of small surprises. The Olympics or the Cascades they are not; but in their absence, Willapa will do.

# Ring of Rivers

Video lurked far in the inscrutable future when I attended Peoria Elementary School in Aurora, Colorado. Our "visual aids" consisted of clunky sixteen-millimeter educational films shown on clunkier big projectors. The films invariably broke or jumped the sprocket, or when they didn't, the bulb burned out. Mostly I remember unrelated flickers on a sleepy screen.

From those pre-mediacenter days, however, two movies stick in my mind. One of them concerned apples in Wenatchee, and I probably remembered it because of its Washington theme. ("Washington" was a magic word in my household. From my grandmother's pioneer teaching days in Sultan and Chelan, and my mother's girlhood in Seattle, a kind of Washingtonian mystique pervaded the maternal side of the family and rained on my receptive consciousness.) Never mind that the story line celebrated that most forgettable of apples, the Red Delicious; I found the images memorable.

The other indelible reel dealt with rain, and the recycling and conservation of water in nature. I suspect that it was a Soil Conservation Service propaganda piece. The simple plot had a raindrop, with facial features, fall from the sky onto a forested hill, irrigate the trees, enter a rivulet, thence a stream, eventually a river, and ultimately the sea before rising to a cloud again as mist. Along the way we shared in all the adventures of Willy Waterdrop—the plunge over a waterfall, the close call with a thirsty cow, the run-in with pollution, and the further insult of having to mix with mud running off an eroded slope. Perhaps the memory embellishes, but I remember the concept of cycle coming home firmly.

Always the animist, I was receptive to a waterdrop's ride and I found the simple story quite impressive. I don't doubt that it had something to do with irrigating the buds of my conservation consciousness. And, of course, it related to my growing infatuation with the family-fabled Northwest, since (I knew) much more rain fell there than we were blessed with in eastern Colorado. Most of all I was intrigued by the actual journey. I envied the drop his ride down the river (he was patently a male waterdrop). I thought I should like to blend with the water that way myself.

I nearly did so when I was three or four. When the clouds broke in great summer thunderstorms, my brother and sister and I loved to walk barefoot in the concrete storm gutters to feel the warm flow on our ankles and smell the sharp scent of rain on the dust. My older siblings recall what I cannot, that on one such outing I fell into the rushing current beside the sidewalk. I was very nearly dragged down the corner drain in the curb, into the storm sewer. I would have fit, too, for those North Denver gutters had wooden lintels fully eight or ten inches high and ungrated. Evidently Susan and Tom each grabbed an ankle as my head and shoulders disappeared, and yanked me back into the light.

Far from becoming frightened of water on that account, I began hanging about the High Line Canal shortly after we moved to the plains suburb called Aurora. That began a love affair with the prairie watercourse that continues today, long after our subdivision on the very edge of the lone prairie has become inner city. For years I wandered the banks of the canal in search of butterflies or whatever else was going. I became interested in its route and obsessed with finding its source and its mouth.

The first was easy: the canal burst from an old railroad tunnel cut through the last crystalline outcrop at the mouth of the Platte River Canyon in the foothills. Thence it flowed in a concrete flume for a mile before assuming its meandering pattern for ninety miles across the high plains. But finding the end of the ditch was something else. I knew the terminus was First Creek (not Kansas, as I had once thought), a small tributary of the South Platte, whence Rockies raindrops returned to the river bound for the Missouri and the Gulf. But try and find it! The canal disappeared into a hydra of small creeks and ditches, none of them clearly its linear descendant. I finally concluded (incorrectly) that the High Line ended in a little reservoir draining into a creeklet tributary to First Creek. Willy Waterdrop, I decided, faced a circuitous route back to the clouds should he fall in the drainage system of my own High Line Canal.

Later, in college, I read Loren Eiseley's *Immense Journey*, wherein he described his own merging with the river—his pantheistic baptism by total immersion in the flow of things. The watercourse he chose for this sensual ceremony was none other than the South Platte. The closest I'd ever come to that sense of watery nirvana was wading in my muddy ditch on a sticky August afternoon. But I got the idea.

Once in Washington, I began spending time around bigger streams. Here the rivers bounded down out of deep green gorges or gushed off glaciers; they cut deep courses that flowed all the time, unlike the ephemeral appearance of the High Line or the mile-wide, inch-deep reputation of the Platte. And as for river mouths, they dumped their loads into big, broad estuaries ringing with littoral resonances: Puget Sound, Nisqually Delta, Gray's Harbor, Willapa Bay—or in the case of the Columbia, the Pacific Ocean.

Nor did the Northwest rivers bear plain names like the Midwest Platte. Rather, they were called by Indian names that flowed off the tongue like the water they bore: Snohomish, Skokomish, Skykomish, and Snoqualmie; Hamma Hamma, Elwha, Duckabush, and Skookumchuck;

Napeequa, Cloquallum, Nehalem, and Chehalis; Queets, Quinalt, Quillayute, and Hoh; Skagit, Stillaguamish, Sammamish, Duwamish, and so on. The rivers of the Northwest are everywhere and dominant. They flow through our lands and lives so that, if we are only aware, we can almost feel something of Eiseley's immersion without ever going near the water.

Willapa has its fair share of these streams. The region may be defined by the rivers that flow around its edges. In the previous essay I described the difficulty of finding a precise geologic basis for this region, since its bedrock runs into that of the Olympics and Cascades beneath a veil of vegetation and recent outbursts of rock and ash. I suggested that we may as well adopt the practical boundaries suggested by the ring of rivers that make this land practically an island.

From its big bend northwest of Portland, Oregon, the great Columbia River flows westward, separating the Willapa Hills from the Oregon Coast Range. Its effluent joins the Pacific, whose big bays may be thought of as tide-rivers forming the western frontier of Willapa. Into the northernmost of these bays, Gray's Harbor, flows the Chehalis River, the natural corridor between the Willapas on the south and the Olympics on the north. And along their eastern hem, keeping out the Cascades except when they erupt, runs the Cowlitz, down into the Columbia. Thus, the ring of rivers.

Let us begin where the sweet water of the Chehalis grows brackish at its outlet and trace this riverine ring around in a clockwise direction. What do these rivers see as they swing around the lump of land we are calling Willapa?

The Chehalis empties into big, spade-shaped Gray's Harbor, where the twin cities of Aberdeen and Hoquiam share the north shore of its estuary. Chehalis water finds its way into the Pacific over the bar at Westport, a salmon-charter and whale-watching town on the tip of a spit. Seaside condominiums fall in line across the harbor's mouth at Ocean Shores.

The bay to their backs is home to mills, industry, and shipping, as well as to hundreds of thousands of shorebirds on the mud flats and salt marshes of Bowerman Basin. Dr. Steve Herman, Evergreen State College biologist, has long worked with others to save Bowerman Basin from industrial expansion. Herman has called the Bowerman birds "a touchstone to a finer time, a richer time beyond our species' memory."

A heretofore neglected wetland stuck between a highway, an airport, and a sewage lagoon, the basin occupies just five hundred acres. As Janet Anthony, organizer of Friends of Bowerman Basin, has pointed out, this remnant constitutes only 2 percent of all intertidal habitat around Gray's Harbor: yet half of the million shorebirds passing through the harbor depend on the basin in spring. Of international migratory importance, Bowerman's fate could redeem or confirm the ecological disasters of the rest of the region.

Following the Chehalis River upstream, east through a broad, rural valley, we pass at Satsop two doomed dinosaurs—twin nuclear cooling towers of the massive mistake known as WPPSS. When the Washington Public Power Supply System planned to build five new nuclear power stations to serve a market that didn't exist with electricity it could never afford to generate, it arranged to service two of the nukes with Chehalis water at Satsop. When WPPSS defaulted on the greatest private debt in fiscal history, the northwest edge of Willapa was spared (for the time being) entry into the nuclear age with all its anguish.

The Chehalis turns southeasterly around the flank of the Black Hills, a small range that separates the northern Willapa drainages from those of southern Puget Sound. Along the old Union Pacific-Northern Pacific railroad line, through the Chehalis Indian Reservation, runs the river that predates them both, and on into the town that bears its name, sister to neighboring Centralia. The stretch prior to Chehalis passes Grand Mound and Ford's Prairie, artifacts of glacial outwash grasslands. These prairies, a rare feature in western Washington, bear the famous and oddly symmetrical Mima Mounds—fascinating hillocks that we owe either to gophers or glaciers.

But what is this? Now the Chehalis, irresolute river, turns west again, its valley forming the route of the main (and only) state highway across the Willapas. It nearly reaches the crest when it gives up and strikes south once more, and that way continues to its headwaters high in the Gray's River Divide. Arising near the Chehalis, the Willapa River runs westward out to Willapa Bay, completing the midrange crossing that the Chehalis began. A third pair of twinned towns, South Bend and Raymond, perch on its outflow.

Actually, the Chehalis is a two-headed stream. The main headwater branches, the South and East forks, bracket Baw Faw Peak between them. Baw Faw, the highest point in the Willapa Hills, rises to 3,114 feet but hardly warrants the title "peak." On its one side, the South Fork flows pastorally through a broad dairy valley with hamlets named Boistfort (the original of Baw Faw) and Wildwood, after sneaking down from its start in the hills. The East Fork arises on the other side of the mountain and is truly a mountain stream. At Fisk Falls, it cascades through deep green gorges of maidenhair and saxifrage, the haunts of dippers and Olympic torrent salamanders. Constricted between stone walls, it runs six feet wide and ten deep in spring, but driftwood wrack high up on mossy bluffs betrays the astonishing depth achieved during winter flood. After the canyon, the green swirls spread into a broad, shallow, and sparkling sheet of stream that runs down to the tired logging town of Pe Ell. A final thrill before the gentle run to the sea, Rainbow Falls carries the East Fork to its confluence with the South.

Just a small eminence separates the Chehalis and Cowlitz rivers. In fact, their tributaries come within a mile of one another at two or three spots in the Cascades foothills. But for small quirks in the countryside, this river that couldn't make up its mind, the Chehalis, might have (a) run around the Black Hills into Puget Sound; (b) joined with the Willapa River not far below their near-common headwaters, for a much shorter trip to the Pacific than it eventually takes; or (c) popped down into the Cowlitz from the Chehalis, then southward to the Columbia. Instead, it twisted stubbornly and persistently around like a clock spring until it forged its own course west, to meet the sea on its own terms. Otherwise, there might have been no Gray's Harbor as we know it, and no cities of Aberdeen and Hoquiam. And I might get my mail addressed to Gray's River instead of Gray's Harbor.

Of course, the river had no such choice in the matter. And the "small quirks in the countryside" that determine its course really consisted of such weighty matters as glacial dams melting and making the Black River as big as the Columbia where it joined the Chehalis; earth uplift on a gradual but decisive scale; and the Gray's River Divide, which looks puny on a map or from a distance, but might as well be the Continental Divide to a river.

Now the Cowlitz River follows a more predictable course, a traditional formula for the successful river. It begins in Mount Rainier National Park, where glaciers nourish its aspirations. Flowing down out of the Cascades, it provides the westerly approaches to White Pass on the Cascade Crest. Midlife crises in the form of big dams interrupt its momentum at Riffe and Mossyrock. But the river regains its equanimity and reaches the valley below. Only now does the Cowlitz become a border of Willapa. Thence, a clear run past Castle Rock, down to its mouth at the predictably twinned cities of Longview and Kelso. These bottomland towns share a bridge across the last stretch of the river, before it gains the Columbia in a miasma of mills.

Before it gets there, some fifteen miles upstream, the Cowlitz takes in the Toutle. With it, in 1980, came thousands of tons of volcanic ash and mud expelled from Mount St. Helens upstream and washed down in the boiling torrent of the Toutle. So now what distinguishes the last few miles of the Cowlitz River, along Interstate 5, is the immense accumulation of this material. Clogging the stream channel, graying even the water itself, but mostly piled into high terraces across the floodplain by derrick and 'dozer, the ash lends an attitude of somber gray to an otherwise green scene. In the winter, when green flees as far as it ever gets and the skies look like the inside of a diving bell; when the new gray range of hills lies stacked between the freeway and a rank of immense red cranes; then the drabs of the ash heaps and the clouds run together to make a melancholic mood second to none I've known.

But even the ash cannot remain barren in a land of little drain. Green fuzz increases annually as mosses, legumes, and course grasses assert themselves. Soon some perennials will get hold, and the gray will begin to give over to green. But the strange new landforms of the piled, planed ash mounds will remain, informing travelers of the essential fact that they exist at the pleasure of the volcano.

Of the ash that ended up neither clogging the Cowlitz nor lining the streets of Yakima or Moses Lake, a great deal washed down into the Columbia River. The Army Corps of Engineers, as if making amends for blocking it with concrete at various junctures upstream, worked overtime with dredges to keep the Columbia open following the eruption. Nevertheless, many ships remained bogged down in the harbors of Astoria, Longview, and Portland for weeks. Such is the central place of the Columbia in Northwest life and commerce that, when it fails to flow properly, life goes on hold for thousands of people and their communities.

The hills don't really care. Volcanic eruptions are a way of long-term life around here, an occasional fact, and the rivers have had to carry heavy gray loads again and again. In glacial times, the river may nearly have failed to flow as ice dams blocked its headwaters; and when they melted, it flowed like an ocean for days, weeks, or months. The hills find their hemlines raised or lowered by all this, and the gradients of their own rivers change, along with the lines of their faces. But they persist much the same when it's all over.

So it is the great Columbia River that forms the third circlet of the ring of rivers around Willapa. Rising in the Cascades of British Columbia, the Columbia rolls on, carving its great coulees in eastern

Washington, defining the northern desert basin of the so-called Inland Empire; ponding sluggish through a dozen dams; trundling on down through its fabled gorge past Portland; and then tidal to the mouth and the bar. It is this last leg, from Longview to Astoria and out over the bar into the Pacific, that forms the southern bound of Willapa.

Sandy islands of poplars and farms dot the Lower Columbia. The biggest of these, Puget Island, acts as if it were a wedge in the river's throat: above Puget Island, the Columbia spans about a mile in width; below, three miles, or five, or more. Buoys show the way for ships between shoals, keeping them in the narrow channel that the dredges constantly labor to keep open. Big river, big ships. Sometimes driving my Honda down the river road, we pass the giant Japanese vessel marked HONDA on its big box sides, and honk; but it never answers. Other ship watchers gaze from the little park at the Cowlitz-Wahkiakum County Line, or from the Vista Park beaches at Skamokawa. The topped white form of Mount St. Helens looks back down the river.

Seals and sturgeon inhabit this river instead of the dippers and trout of the Chehalis. Its tide runs up on big sandy beaches made of pumice and dredge spoils. Yet the little rivers with their eddies lapping at pebble shores have this in common with the Columbia: they drain the green sponge. They allow the land to bear the burden of four meters of rainfall. The drippings of the mist run between their banks. Rain into green, rain into soggy soil, rain into rivers. Rainmade rivers define the landscape and lifestyle of the region, just as they determined its pattern of exploration and settlement. And like Lewis and Clark, the river itself finally reaches the sea.

So the ring has the blue Pacific as the gem on its western setting. Offshore, the Japanese Current, a river in the sea, could be the western edge. But the shoreline, and the Long Beach Peninsula, have little to do with the Willapa Hills. So I recognize Willapa Bay, where riverine becomes estuarine and fresh water meets salt, as the ring of rivers' finishing arc. Complex system of its own, Willapa Bay receives mere mention here.

Willapa, surrounded by its watery circuit of streams and sea bay, is nearly an island. A short canal between the Cowlitz and the Chehalis would make it one. But those boundary rivers are hardly the only watercourses of importance in the region. A pattern of streams and creeks, sloughs and brooks, dissects the green land, the better to receive their runoff. I've already mentioned the Willapa River, which rises in the hills near the Chehalis and eventually becomes its parallel counterpart draining to the west, with half a range of hills between them. Gray's River also rises near these two, separated from them by the Gray's River Divide, and it flows south into the Columbia at Gray's Bay.

Other streams flow here, among them the Naselle, Deep, Chinook, Bear, Palix, Bone, Nemah, Niawiakum, North, and Elochoman rivers; Teal, Seal, Steamboat, Brooks, and Greenhead sloughs; Skamokawa and Salmon creeks; and many more. The names reflect the Indian occupation and the English period of exploration that preceded the Scandinavian era of settlement. Altogether, the rivers define the life-patterns of the Willapas. Transport and commerce, the distribution of plants and animals both over time and within their own lifetimes, the drainage of rainwater, the distribution of floodwater, the building up of the valleys and the breaking down of the hills, all depend upon the rivers' flow.

The rivers have determined where people could travel and settle, farm and fish. Today highways having taken over travel routes, the rivers mean mostly fish. Commercial fishing has always been a major source of income in the area, and though much reduced, the fleets still go out. Sport fishing brings more visitors than anything but clamming and hunting. Along the Columbia at strategic places, one sees the sturgeon fishermen encamped in their aluminum wombs, as their Indian predecessors would have been in hides and brush. Offshore in the tide-running river, gill netters pull in their great baskets of salmon as smelters dip their long nets into the swarming schools. Throughout the wintertime, fishermen who love their sport or quarry more than comfort line the Gray's, Elochoman, and Naselle rivers in pursuit of steelhead, the anadromous (seagoing) version of rainbow trout. In lesser numbers, birders, canoers, and pleasure crafters, mostly from the cities, work the curves of the rivers and sloughs for their own kinds of rewards. In such ways, the people stay in touch with the rivers.

That's when the rivers work. What about when rivers go wrong? As the blood flow of the land, the rivers are bound to suffer from abuses to the body of the land. We have seen that when its channel filled with volcanic ash and mudflows, the Columbia did not work as it is expected to do. The Toutle and Cowlitz still carry so much effluvium from the recent vulcanism that their ability to hold a flood has diminished tremendously. Until the recent completion of a tunnel to bring down the level of Spirit Lake by Mount St. Helens, riverside residents quaked when it rained, awaiting the inundation of several towns that an overflow could have brought down the river.

The power of rivers to move materials to the great septic system of the sea has caused them always to be used as sewers. Governor Tom McCall showed how the Willamette River of Oregon could be cleansed, and discharges everywhere began to be monitored more carefully. Yet the Clean Water Act stands under attack in the environmentally moribund Reagan administration, and the Hanford Nuclear Reservation in eastern Washington has been named as one of three candidates for a national high-level radioactive waste storage site. The shafts, dug deep into porous flow basalts along the Columbia, would very likely drain into the river. A café wag quipped that at least sturgeon fishermen could abandon their lanterns: the fish would glow in the dark waters.

On March 19, 1984, the tanker *Mobiloil* ran aground near St. Helens, Oregon, a river town with a grand view of what's left of the mountain. At least forty-two thousand gallons of heavy industrial-grade oil poured into the Columbia River. Thousands of western grebes, murres, scoters, and other sea and river birds took on a lethal coating. A rescue center was set up in a Fish and Wildlife Service facility near Skamokawa, so that when I captured an oiled murre in an insect net on a beach at river's mouth, I had somewhere to take it for attention. As "my" wildeyed bird was logged in, given a high-protein drink, and set into a heated cage, I scanned the rows of such boxes full of desperate creatures. All oil-industry people, executives and supernumeraries, drillers and drivers and skippers and pumpers and stockholders, should be required to visit such a center. All accessories to a spill should be sentenced to work beside the volunteers who clean up their messes and try to quiet and comfort the terrified birds.

Thanks to such volunteers, hundreds of the besmirched birds were saved. Many more died of toxicity and exposure. Wildlife agents feared for otters and other furbearers as they plied the intertidal zone, where oil globules tend to aggregate. I think of *The Wind in the Willows* and Kenneth Grahame's ambivalence over the changes brought to "sweet Thames" by industry; and what these changes might mean for Ratty and his other small river friends "that glide in grasses and rubble of woody wreck." Newspapers raised concern for fisheries as a result of the spill, the cleaned birds were released upcoast to catch up with their comigrants, and we heard nothing more about it—the way of all press.

Things do go wrong with rivers, and much of what could go amiss with the Columbia already has. A full deck of dams has converted its biggest tributary, the Snake River, into "Snake Lake," and the mother river into a string of placid, lapping ponds. The great cascades where the Indians netted salmon have all been inundated, along with most of the habitat of the Columbian white-tailed deer, the Columbia tiger beetle, the Oregon swallowtail, Shepard's parnassian, and many other organisms, large and small. This has not been accomplished without enormous losses to the anadromous fisheries, for all the fish ladders and lifts in the world have not begun to compensate for these massive frustrations to the wanderlust of the salmon.

Only two stretches of the Columbia in Washington remain free-flowing: the white-cliffed section due for stuffing with nuclear wastes near Hanford and the tidal portion that borders Willapa. It's a little late, but finally the West's greatest river has ranks of supporters girded to protect the last uncompromised values. The nuclear-waste dump may never come about, so vocal and convincing are its detractors: only a triplet of towns, whose job base has depended largely on atom-busting ever since the Manhattan Project, seems to want it. Any serious move to build the proposed Ben Franklin Dam, thus making slack water out of the last free-flowing forty miles above tidewater, would find many bodies arrayed before the bulldozers.

The spectacular Columbia Gorge, already compromised in many ways, finds itself the subject of energetic lobbying for federal protection of what's left. Only a national scenic area or recreation area under the management of the federal government has a chance of saving the gorge from the vandalism of piecemeal development: so say gorge-protection activists. Local governments and their backers think they can do a better job of taking care of their backyard. It's the oldest story in the conservation book. Judging from the way many locally managed attractions have been treated—the Wisconsin Dells, the Royal Gorge, the private precincts surrounding several natural parks, for example—I believe the federalists clearly have the better case. If they can also win in Congress, the gorge may yet remain capable of giving meaning to that devalued word "gorgeous."

The Lower Columbia lies outside the gorge proposals, although those of us who travel it frequently know that many of the same charms can still be found well downstream from Portland, the official end of the Columbia Gorge. The importance of this ultimate portion of the river was emphasized in 1977 with the formation of CREST, the Columbia River Estuary Study Taskforce. CREST seeks to provide a basis for protecting the natural values of the estuary, while recognizing and facilitating its enormous economic role for the region. Such overtures give one cause to hope that the river may not be taken for granted in the future the way it always has been before.

Remarkably, most of the rivers of the Willapas remain wild—at least in the sense of being undammed. They are short rivers running down steep ravines, perhaps not on the whole presenting good reservoir sites. Or perhaps the need has evaded even the slender case for cause required by the Army Corps of Engineers or the Bureau of Reclamation. In any case, one rarely finds so many streams in proximity, none of them having suffered dams. True, the Cowlitz has dams upstream in the Cascades, and of course the Columbia. But on the rivers arising in the Willapas, we do not find significant impoundments.

The bad news is that they do not, on that account, necessarily flow free. Natural debris in old-growth forest streams provides aquatic habitat and stream nutrients. But clearcut operations frequently fill adjacent streambeds with vast quantities of slash-limbs, boughs, bark, and small trees not worth the taking. Slash-damming can impede streamflow as much as dams, while adding few of the benefits of large rotting logs.

Between slash, overheating from lack of shade, and siltation from erosion, fish have little chance to spawn successfully. Although loggers and commercial fishermen are not frequently the same, or even of the same families, nonetheless they share in a common economy. It always surprises me to see the slash of a clearcut piled into a stream like the cobbles left behind by hydraulic mining in a Colorado gold-rush stream.

Today, greater care is said to be taken. I don't believe it. The Forest Practice Act basically says "do this and that to protect the streams; but don't bother if it's too much trouble or expense." Slash-damming and erosion into rivers continue dramatically in these hills. Of course, carrying silt to the sea is the perpetual job of rivers.

But in the quantities released by steep-slope 'dozing and logging, the rivers can't cope, at least, the life in the rivers can't cope. Their populations drop like the tide when their feeding or spawning grounds get dumped on.

According to that grade-school movie, this wasn't supposed to happen. If it did, the soil, and the rivers, were in trouble. And so they are. In flood, the ring of rivers runs brown very quickly. Upslope, the soil simply isn't being held as it should be. So erosion may happen faster than it should, and we will pay. I have been informed by a county commissioner who is also a logger that there is no erosion in these hills. But something brown fills the rising river, and it sure looks like dirt to me.

Erosion: rivers forming and giving form to the land. The ancient river terrace on which I live bears witness to the process. When the rains come hard and Gray's River floods the valley below, the upslope stone walls of my cellar leak like punctured buckets. And whenever I look out my windows at the river, I see the movement of water down the valley. These things link me to the riverworld. Sometimes that old documentary film comes to mind as I watch the river. It occurs to me that Willy Waterdrop might be out there now, one more time making his way through the rain cycle, part of the constant flow of water from cloud to earth, round the ring of rivers, and down to the sea.

# Robert Gray's River

The Columbia River spent a long time not being "discovered" by Europeans. Fabled for many years as the speculative River Oregon, its fierce tidal bar kept would-be penetrants at sea. In 1775 the Spanish mariner Bruno de Heceta sailed past, took one look at the bar, and decided to give it a miss. He called the southern bluff Cabo Frondoso for its lush vegetation; the northern, Cabo San Roque; and the forbidding river mouth, Entrada de Ezeda. Mapmakers in Madrid inferred a river from his charts but dropped his name, dubbing it Rio San Roque.

Next came Captain James Cook. Seeking a Northwest Passage at the sixty-fifth parallel or higher, he stuck to his orders not to detour into any bays or rivers south of there, and sailed right by in 1778. The next British seafarer in the region, Captain John Meares, was a skeptic regarding the putative St. Roc River. In 1788, he weighed anchor off the bar, examined it, and concluded there was no river. He renamed Cabo San Roque as Cape Disappointment and the estuary, Deception Bay.

George Vancouver, too, perhaps wishing to avoid disappointment and deception, sailed clear past the river mouth in *Discovery* in 1792. He reported that no safe harbor or major river existed between Cape Mendocino in northern California and Cape Flattery in northern Washington. Everyone knows Vancouver, the noted British explorer, for the great Canadian city and island in British Columbia that bear his name. Less well known is Vancouver, Washington—a twin city of Portland, Oregon, across the Columbia River from it. But if Vancouver ignored the Columbia, how came his name to be applied to a point, then a fort, finally a city some ninety miles upstream from the river's mouth in Washington State?

It happened because Captain Vancouver had a change of mind and dispatched Lieutenant Broughton in HMS *Chatham* to chart and explore the Columbia. Broughton did so, naming many features along the lower stretches of the river, including Mount Hood and Point Vancouver. So it is that we might well have a Vancouver River to further confuse Northwest geography, already confounded by Vancouver, Canada, and Vancouver, U.S.A. In fact, had the Englishman Broughton been the first to sail up the Columbia, there might well be no Vancouver, U.S.A.—for Washington and Oregon and perhaps a good deal more might have become provinces of Canada.

But Vancouver's man was not the first. In between Vancouver's initial miss and second shot, one Robert Gray of New England slid over the bar and "discovered" the Columbia River, in May of 1792. Captain Robert Gray, a Rhode Islander, skippered the ship *Columbia Rediviva* for a Boston trading company owned by Messrs. Barrell and Bullfinch. He plied Northwest waters in search of sea-otter skins, for which he traded copper to the Indians. He took the furs to China to trade for goods to be returned to his employers in Boston, becoming along the way the first American mariner to circumnavigate the globe. On the way, he accepted the challenge of the thundering bar and found "a noble river," which he named for his ship. Thus were George Vancouver's hopes of first finding such a river and claiming its territories for the Crown frustrated by a Yankee fur trader.

As it happened, *Columbia* and *Discovery* met on the high seas off the north Washington coast after Gray's initial failure to reach the river. His negative report to Vancouver reinforced the British captain's skepticism.

Nine days later, on May 11, 1792, Gray's thirty-seventh birthday, he tried again and succeeded in crossing the treacherous bar. Eight or nine days elapsed as Gray traded and at one point clashed with the Indians of the Lower Columbia. The ship went aground in a muddy bay on the north shore, washed off with the tide, and sailed in stages back downstream. Dramatically, Gray encountered Vancouver again at the Spanish settlement on Nootka Sound, Vancouver Island, the following September. He confided his discovery to the nonplussed Vancouver. Soon thereafter, Vancouver sailed to the mouth, armed with Gray's crude chart. Unwilling to risk *Discovery* on the bar, he sent Broughton in the lighter *Chatham* across. Broughton proceeded to claim all the lands drained by the river for the King and later claims were made that Gray never really entered the river, having remained within the tidal mouth. But the commissions that decided imperial matters disallowed Vancouver's priority, and the border was fixed at forty-nine degrees. An exception was made for Vancouver Island, the Canadian boundary swooping down and out the Straits of Georgia and Juan de Fuca so as to include the big island in the Commonwealth. But if Vancouver got his island, Robert Gray kept his river.

So far, one might think that the title of this chapter refers to the Columbia. Indeed, for the crucial crossing into it Gray (had he been a less modest man) might well have named the big river after himself. As it happened, another river bears Captain Gray's name, a much lesser stream yet one not without its charm and interest. For that shallow-water place where the *Columbia* temporarily went aground received the name Gray's Bay; and subsequently, someone called the larger of two rivers flowing into it Gray's River. This river's outlet clearly shows on Gray's rough chart, and it is here that the American captain seems to have gone ashore briefly during his adventure up the Columbia. So the name fits, and it sticks. It is that little river that I write about now.

One of its chief blessings for me is its obscurity; still it is a pity in a way that more people don't know Robert Gray's river. It is worth knowing. Headwaters lost in foggy canyons, outflow in an estuarine bay of the Columbia, Gray's River furnishes a metaphor for the region. Its slopes and ravines logged, its riffles fished, its fertile floodplains farmed, and its tidal reaches once plied by tall ships and packet steamers, Gray's River flows like the lifeblood of the community that bears its name, where I live.

From the village of Gray's River, one looks northwesterly into the higher slopes of the Willapa Hills. There the stream arises, near the headwaters of the Chehalis and Willapa rivers. The hundred inches or so of rainfall allocated to the hills sorts itself into these three and other basins with the help of the Gray's River Divide, Huckleberry Ridge, Long Ridge, and other landforms. You can't reach the very sources without scrambling across acres of clearcuts well beyond the logging roads. But by driving up onto the Divide on a weekend, when the log trucks are idle, you can get a look at the upper stretches and see about where they come from. First, it is helpful to look at a map.

When I outline Gray's River, its forks and tribs in yellow on a detailed map, I get a dendritic pattern. More than an elm, it resembles some hedgerow shrub left untended. A squiggly brier bush of a river, its branches bend at the least contortion of the land. I have made no attempt to calculate the length of the many watercourses making up the pattern of the whole. But in the Wahkiakum County Comprehensive Plan, I read that the Gray's River basin incorporates 124 square miles of land, or 79,400 acres. A great deal of rain falls on that much land in a year, and it is the river's job—and that of all its tributaries—to move the runoff down to the bay.

There seem to be three main trunks: a discrete, short West Fork, running up past a salmon hatchery to the high slopes of rugged K. O. Peak; an even more distinct South Fork, flowing out of the east and separated from the other rivers of Wahkiakum County by the southern arm of the Gray's River Divide; and a main branch running more or less north-south through the center of the Willapa Hills. Long Ridge splits it into the East and North forks, while the northern Gray's River Divide keeps it out of the Willapa River watershed. Once more I am struck by the chance nature of uplift and erosion in determining a water drop's destination: at various points less than a mile separates the West Fork from the Naselle River; the North Fork from the Willapa; the East Fork from the Chehalis; and the South Fork from the Elochoman.

Since I have been living in Willapa, I have explored twigs, branches, and trunk of this scraggly water-tree, on foot, by canoe, and on logging roads by auto. The logging roads penetrate almost every stream's upper parts, giving rough, rocky access to the backwoods, although often there are no woods when you get there. Let us begin at the crest of the Gray's River Divide, northern version, and descend with the river to its languid outlet in Gray's Bay.

Remember that the Willapas are a range of ridges and ravines. On the roof of the range, the rivers drop out of the heads of ravines as mere rivulets. Since the Divide supports no permanent or even winter-long snowcap, these rivulets arise from springs and swell with precipitation. But at their very outset, they are as likely to be quick, tiny waterfalls down the slopes of stony seeps as sluggish puddles issuing from mossy crotches way up the steep ravines. You would scarcely recognize a river in these high-level drips.

Surprisingly soon, real streams collect from the numerous streamlets, giving mapmakers something to call the river. These channels quickly gain a yard or two in width and a foot or two in depth (in the deeper holes, in winter) as they pierce the salmonberry thickets of the high, steep slopes. Then as they reach the greater folds they find one another and coalesce into creeks—Cabin Creek, Alder Creek, Blaney Creek, Beaver Creek—mostly permanent streams that tumble over their stones as if they can't wait to reach the gathering stem of the river itself.

When they do, it's already a river, ready to take on reinforcements. At a point just eight or ten miles downstream of the highest headwaters, after the North and East forks have met and decided to team up but before recruitment of the South and West forks, Gray's River already spans twenty or thirty feet. Though broad, it runs fast and shallow like the upper Wye on the English-Welsh border, with a constant whooshing sound that reminds me of the before-storm voice of the wind on the High Line Canal in Colorado.

Making such comparisons, I am struck by how we—or I—always tend to do so: this place reminds me of that place, and that one of this. How, then, is one to appreciate a place on its own merits? Perhaps we do this out of a wish to combine all pretty-good places into a perfect place in our minds. For example, the Wye has charming villages and inns and a vast bookstore in a ruined castle, but lacks wildness, tangle, and bears. The High Line Canal possesses the sweet associations of youth when first on its own in the out-of-doors, and the sweeter scent of the gray willows and green cottonwoods, but its scene bears no relief, no up and down. And here is Gray's River—wild, all up and down, the possibility of bears. But there is no soft human touch of hedgerow, stone wall, or village, as along the Wye; no innocence of view such as a seven-year-old might possess beside a prairie ditch in 1954. The human touch comes in clearcuts, which do away with any innocence at all.

The valley widens a little between the hills, and here the scene is one of beauty if you look at the river and its banksides alone. Alders and maples line the shores, grassy banks run between. A great mossy maple overhangs the broad, open river, dripping licorice ferns off its moss-furred branches in bright green plumes. In the stream, a pair of dippers bobs on mossy stones and jabs the eddies for food.

Then you look up and see the savaged steep mountains all around. Steep, steep slopes have all been clearcut and have eroded desperately. Some of the slopes have been replanted, and the winter huckleberry's red contrasts nicely with the soft green of the ranks of young conifers. But you wonder, what chance for a big forest ever again on these slipping, eroded slopes? And what of the river life, as that soil rushes downstream? Without forgetting these things and their lessons, you must temporarily edit them, if you are to enjoy the remaining beauty of the basin.

Now the river passes its halfway point in terms of elevation. Beginning at around two thousand feet, ending at sea level, Gray's never knows the three-mile drop of a Platte or the two-mile plummet of a Rainier glacial outpouring. The cutting power of a river is determined by its gradient. The Gray's may not arise at great height, but it is also not very long. Its gradient, therefore, remains considerable until it reaches its floodplain. This river cuts out the hills' features still.

Now the South Fork comes in from the east, and the collective river rounds the flank of the southern embodiment of the Gray's River Divide. We cross a bridge deep in second-growth forest but still high above the river. The clear-cuts have retreated upslope and can be edited with less obvious effort. The river shows its gradient as it tumbles and roils over the moss-softened, river-rounded stones of its bed. Here the water runs black and green and white-black where it pulls deeply down its course, white where it strikes stones and flashes back on itself, green where emerald algae show through as streamers in the flow.

One imagines the powerful salmon working their way against that strong current, showering through the water-flashes, sliding off the seaweeds. Dark, deep pools hang on the lee sides of boulders, where slide-tired salmon might rest before the next upward plunge. Anglers know this and find their way somehow, down through alder brake and spruce fence to the rocky shores. The river here almost has a roof, made of branches hanging over the flow, all festooned with club mosses and long chartreuse lichens that sometimes reach the water and flow along with it. When I worked my way down into this true mountain gorge, I found splash-pools inhabited by newts and frogs, caddis flies and water striders. All was green wilderness and rushing water and whistling dippers, and the logging beyond could be forgotten.

I used to imagine that no one knew the looks of the whitewater stretch below, unless it were some intrepid kayaker. A naive, ironic assumption, that: it's the very part the Public Utility District has proposed damming for hydro-power. The motives, to lower rates, gain local control over power production, and get out of the Bonneville Power Administration's nuclear grid, are noble ones. But for a river to escape unimpounded this long only to be dammed in its best stretch at this late date would be more than regrettable. The politics, power surplus, conflicts with fisheries, and additional permit hurdles will probably prevent any dam on the Gray's. I know I, for one, would fight it.

Each of the bigger rivers of these hills has wild miles of green gorge walls and blue water-scooped holes. On the Upper Naselle, I like to look down on a suspension foot-bridge between two old homesteads, where the river tumbles below in white splashes. A little farther up, turquoise trout-waters carve sandstone bedrock into shell shapes beneath miraculously uncut old-growth spruces. And farther up yet, a big, cool eddy known as the Lahti Hole boils darkly beneath a wall of ferns. Salmon Creek, between the Naselle and the Gray's, has a hidden falls beside it that no one would see who did not prowl the hills for timber, game, or visions.

Anyone, however, can drive up the Fossil Creek Road to the bridge I speak of on the Gray's and admire the surprising scene of a briefly wild river in the tame Willapa Hills. The new bridge is concrete, with low sides—evidently, log trucks don't need guardrails. But the old bridge, or what is left of it, will leave an impression on the visitor as memorable as the river itself. The span itself is long gone. But its buttresses remain, and they are amazing: seven-layer log sandwiches of old-growth timber jammed into the bank so as to raise the crossing to the level of the canyon walls. Two huge logs lie perpendicular to the river, then three at right angles to the first layer, in a gargantuan crosshatch repeated again and again, with earth stuffed in between to make a solid foundation. Seventeen Brobdingnagian Lincoln Logs in all, Douglas-firs measuring up to five feet in diameter. They give the least hint of the size and abundance of the trees that once were. In their architectural entombment, these bridge abutments preserve some of the few logs from the first cut to survive anywhere. Perhaps they will be the petrified forests of the future.

From the road, below the bridge, the river comes into view from time to time. Without quite falling or cascading, it drops several hundred feet in a series of downhill swings. Fossil Creek pours in from the east. The name refers to a relative abundance of paleontological relics in the shaly outcrops broken by the road and its borrow pits. Denizens of quiet waters, these include clams and snails in shale beds and crabs in concretions. The shoreline orientation of these communities, some tens of millions of years old, shows up in the maplelike leaves interbedded with the marine invertebrates. If the presence of marine crabs and clams in the hills seems outrageous, consider that sea level lies today just a short distance down the road. And if maple leaves and sea snails seem strange fossil-bed fellows, carry on with me in this vicarious voyage downriver.

The road leaves the mountains and strikes the valley. Here we leave it and continue on the river itself. Having lost most of its elevation by the time it joins the West Fork, the river too leaves the road. It bubbles past a point where the water supply for the town of Gray's River is drawn off, a well bored into the bedrock beneath the river. Then comes a big gravel bar known as Gorley's Beach, the starting point for the annual event known as the "Gray's River Not-Quite-White Water River Run." Founded in 1969 by Phillip Raistakka, the race covers the final five miles of the river above its tidal reach. Classes include every sort of vessel from kayaks and canoes through inflatables and rowboats to anything that floats.

After years of watching from my veranda and knowing only bits of that stretch of river alongside of which I reside, I entered the one-man canoe class last June 16. Manhandling my eighteen-foot Oldtown fiberglass canoe, *Ms. Wahkiakum*, into the water at Gorley's Beach, I took off among an armada of "Jaws" rafts, bathtubs, and innertube creations. The current, fresh from the rains of a gray month of May, took us away. Here is what I saw.

Bucolic in the extreme, the river passed first between soft green banks lined with alders trailing lichen tresses, buttercups speckling the meadows at their feet. Daisy-spattered pastures ran down to the river to drink along with the scattered herds of Holsteins they nourished. Such a scene dulled me into an unwatchful state when I noticed other craft going ashore and people portaging. The reason: an acute river curve with a proper rapid going into it and a great, swirling rush coming out—a patch of *quite* white water! A bystander positioned to pull out dunkees poised himself, as watchers shouted phrases of mixed encouragement and dire prediction of imminent capsize. I took the curve, cut hard into the eddy, and made it around right side up. None of the rest of the run proved quite so white as that, but I kept alert for further instances of false advertising just the same.

The river leveled and meandered now between family groups on the strand and lively watchers on the highway bridge before entering another quiet stretch that might have been drawn from any pastoral country in any century. A fisherman's shack stood on the shore beneath ancient alders. A long rope hung from a moss-matted maple on the opposite shore, for a different kind of play. The rope indicated a deep hole for swingers and swimmers, but on my side I could see gravel and then dragged keel. I had to climb out and pull the canoe, and then I knew I'd been right never to attempt the river run during a normal low-water June.

Thankful for once for the rainy month of May just past, I shoved off into the current again and soon overtook my neighbor Joel Fitts. The big, jovial dairy farmer skippered a raft covered with 4-H kids as thick as mice on a mousetrap in a flooded basement. Evidently drinking beer and making people wet are as much a part of the river run as running the river. Joel's was one of the few "dry" vessels on the run. But his frisky lot wouldn't let me get by dry, and they splashed until I dripped. I swore revenge.

Again I had to tow the craft as I came in sight of the long, gray covered bridge—the last in Washington and a daily fixture in my life.

Then there I was, afloat beneath it, and in sight of my own house on the hill above. How many times I had watched the water flow beneath that bridge, watched boaters, logs in flood, mergansers in the fall, or just the glitters of the river, from my porch or bedroom window! Now, for the first time, I was looking back up at the old white house, myself a part of the river flowing.

The next section, familiar to me on foot, I had to myself. Spotted sandpipers skittered and piped over the sandbars. A belted kingfisher cut across like a blue rattle thrown in anger. Out of a moss-capped stump sprouted a bouquet of spring beauty, and from the foot of a silted piling a particularly dense spray of goatsbeard grew, French vanilla against the green-black bank. Red, soft soil strata lay behind it, as poorly lithified as last week's pancakes still on the platter. The thought that I was likely the only boater to notice all these features struck me as supercilious; then, on reflection, only realistic.

But the big groups of onlookers splayed on lawn chairs and football blankets on Torppa's and Badger's beaches—each few clustered around a cooler, more people together than one ever sees in Gray's River or even supposes to reside there—hailed me in friendly fashion. They also roared when the wind came up, caught my unballasted bow like a sail, and swung me round 180 degrees from the finish line: the better to watch the wildflowers, I figured, but it was hard work getting pointed downstream again. Rounding a rich meadow bank beside old rotted pilings, I enjoyed the pleasant sensation of a delinquent pesticide odor being overwhelmed by the sweet scent of a copse of wild roses.

Now I swung around a section of loop that I'd frequently run, walked, cycled, or driven. It brought me to the last bridge of the course. A four-man canoe in front of me attracted most of a volley of water balloons launched from the high span, but I caught two between the shoulder blades. The river-run crowd drinks hard, plays hard, and aims to maim. Perhaps it had inflicted the damage that led to the floundering of a raft of Coast Guardsmen wearing bright red survival suits. I passed them pulling one another out of the last deep serpentine of the stream before it enters the hamlet of Gray's River. Here the flow becomes tidal, and here the river run ends. The race was run in one hour and twenty-six minutes. I won my class, which I learned had been canceled for want of entries.

Loath to leave the river as the sea mist fled and the sun came out, I basked across the stern with my feet up on the gunnels and had my one beer, saved until now. One of the most abstemious men on the river, I watched in pleasurable repose as the motley flotilla, or word of its sinking, came in. As I floated in the boat basin, beneath banks once graced by stores, a hotel, and a whitewashed, steepled church, a noisy, beery crowd celebrated on the deck of the tavern above. Such changes this river has seen! A culture come and largely gone in the space of a century. Where once packet steamers arrived daily from Astoria and a hundred fine wooden craft might gather for a big event, rubber rafts and Bud are the best we can do today. There is life in the lowlands yet, but the heyday is done. The river silts up and could no more carry a steamer than a submarine, should steamers still exist and decide to call on the wharf once again. The river run is the only time that more than a boat or two might be seen together here any longer, except when the quirky smelt run up Gray's River of a rare spring such as this just past.

Watching the brief ceremony, I thought about the mortality of towns and rivers. Here is a community where, I am often told, "there's a heck of a lot of history." It's true; a century plus has passed since the first settler, a man named Walker, came. That time has seen a great richness of events, endeavors, and accomplishments in Gray's River. But what is left now is mostly just that—a lot of history. The richness that remains lies in survivors, stories, and a soft acceptance of change. And in the river—which, though mortal like towns, may flow a good deal longer than any of us will be here to see.

At last Joel's raft arrived. Eager to return the favor of the thorough wetting I'd received, I leapt onto the raft and attempted to shove Joel overboard. But my broadside failed: I merely bounced off the massive man into the river myself. Later I sneaked back with a bucket and evened the score. Bob Torppa, county commissioner, finish-line judge, and timer, ruled it a draw and we all repaired to the Grange Hall for awards.

So this is the peopled stretch of the river. Farms, barns, houses, bridges, a very small town consisting of a café, a boarded-up blacksmith shop, a church or two, the post office, gas station, fire station, a handful of homes, the Grange Hall, and the tavern busting out with river-run revelers—that's about it. And this is downtown: downstream, the action thins out, and the shore shows many gray, old homesteads rotting down to nothing, in between pastures, woods, and the few still-occupied farmhouses.

Still, these last few miles of Gray's River, as it rises and falls to the tidal moods of the Columbia, are among its most attractive. Unlike the river-run course, paddled just that once on my part, the waters of the lower reach have parted before my canoe many times. A mile or two below the Gray's River townsite lies an even sparser village. Rosburg claims a store, an attractive brick-and-clapboard school, a cemetery, a tidy dairy farm, and a village hall. Its post office serves a small and dispersed population from Rosburg to Eden Valley to Pillar Rock. Here a gracefully arched white timber bridge, which carries the road to the abandoned canneries of Altoona, mirrors the curve of an old barn's roofline. Just below the bridge may be found a convenient boat landing, and it is here that I like to put in to explore the lower Gray's River.

Now much broader to accommodate the twice-daily overflow from below, it looks like a real river. For a while it lies like a basking blue snake looped across broad old bottoms farmed for their forage, silage, and hay. When I view such a generosity of grass, I think of some vegetarians' argument for converting pasture to arable farmland. More protein can be made from millet or soybeans, goes the riff, than from stock fed on the same land. There may be some very sound arguments for vegetarianism, but this is not one of them—at least not for here.

I think of Britain in the Common Market, forced by the Gnomes of Brussels to convert old, flower-rich grasslands into sterile, blowing grainfields, ripping out hedgerows in the process to accommodate the big new metric combines. Loss of natural diversity and classic countryside result, and grain mountains simply replace the butter mountains. The fact is, Britain grows grass best. The same is true of these Wahkiakum valley farms—whether for burgers or milk shakes, cows grow well here on the rich green grass. The decline of the once-great local creameries cannot be blamed on the grass, that much is sure. The markets have moved. But they would not return for soybeans, that's just as sure.

Floating downstream, we pass the doorsteps of elegant farmhouses, long since abandoned to the freshets and the mice. The backs of old barns break and ancient boats and Studebakers deliquesce into the fundament. Rot comes quick to the objects of neglect in the rain world. But not all here is death, by a long shot. We share the stream with other ripplemakers: otters, coypu, muskrats, and mink. The air overhead, almost always blue on days we choose to canoe, splits to admit the chittering forms of ospreys and eagles and gulls.

Now the ground grows waterlogged and swamps close in, and dikes appear to hold in the fields. Seal Slough glides off to the right like some mysterious bayou. Dense vegetation begins to crowd the riverbanks. Great, leaning Sitka spruces loom over the river, blue presences from some former age. The dairy cattle and their pastures came only recently to Gray's River, but the spruces have loomed this way as long as we can imagine. Their needled branches shade the muddy recesses among tidestained roots, both holding dark secrets in nest and holt and shadow. A tatty fabric stuck up by steel pins, pale green skeins of lichen dangle from the boughs.

Beneath and between the massive spruces, native shrubbery binds the riverside rampart into a variegated wall. Lined with osier dogwood and ninebark, the banks change colors with the seasons. They snow the lush jungle with their white flower-clusters in June, then bloody the river margin with October reds. Tall orchids like white pillars clump around the landings, adding to the summer snowstorm. In the fall, their seedpods and those of irises punctuate the banks with their odd brown forms. Then next spring, the iris comes on in stacks of green spears and brilliant yellow clusters, rivaled in brightness only by the summer clumps of golden lotus that crowd the canary grass or crown the rotting old pilings, each a dense and unique island of life.

Here in the lower river, we have a feeling of great fecundity, as if we have suddenly paddled into Yucatán, or the Jurassic. I find it thrilling, this ragged prolificness, though some from tidier places see it as sinister.

Drifting with the tide or lolling against it, I wish I could spend forever in this way. But it cannot go on. Clear days often cost cool nights, and we feel like pulling out and hauling home. Or a squall comes up, whipping the river into an unwieldy waterslide and putting us ashore. Or we simply come to the end. Scarcely thirty winding miles from its sources, Gray's River reaches Gray's Bay. Here, the Columbia flows its broadest. Oregon rises on the other side, sandy shoals and islands sprawl in between. On the left as you enter the bay, stone-cropped cliffs run around to the cannery ghost towns upstream. On the right, a firred toe known as Miller Point separates the mouths of Gray's and Deep rivers. Between them spreads a marshy broad of offshore shallows loosely vegetated with emergents.

Clumps of pale mauve asters daub September's estuary, when red admirals are likely to be seen cutting over and touching down to drink. April brings a rare marsh marigold blooming yellow among the tide wrack, and another surprise—vegetating *Veratrum*, the lush corn lily, something (like marsh marigolds) that I expect to find in mountain heights. On one expedition, my family and I landed the canoe on a loggy point on the edge of the bay to picnic and peer behind the curtain of vegetation; the marigolds sprinkled grassy islets. Another time, my friend Fayette Krause and I kept to the canoe and crossed the reedy estuary. Two sora rails burst from their hideouts and shot across our bow, black pellets more running on water than flying and so stretched out that they clarified the idiom "thin as a rail."

On the eastern shore of Gray's Bay, a gray clay slope rises from the waterside road. Here, band-tailed pigeons come for minerals, giving it the name Pigeon Bluff. The dove-colored cliffs slough down onto the shore from time to time, stopping the school bus, bringing out road crews to clear the highway, and creating tomorrow's bay-bottoms. Onto the clay fall maple leaves, which mix with mollusks in the tidal shallows. In the formation of these contemporary strata, we can imagine the fossils of the future: maples and snails together, as in the enigmatic deposits we passed upstream.

Toss in the terrestrial snails that frequent these bluffs and the alligator lizards that empty their shells, as well as a lurking sturgeon in the bay, and add an age's uplift: you will have a mulligan stew of a fossil bed to challenge the paleontologists and delight the creationists, should their benighted numbers survive the epochs. And so, land and water creatures come together, as rivers mix, in Robert Gray's bay. The centuries blend as well. Here it was that Gray himself came up and went aground on those same clays. The keel of *Columbia* scratched the mud where I pull my canoe ashore today.

A local tradition has it that Robert Gray took a lifeboat all the way up the tidal stream to the present-day townsite of Gray's River. He would have seen no whitewashed church, no tavern, not even the great elms that today grace the lonely tidal basin. But it is pleasant to think that the town's namesake first laid European eyes on the site of future settlement.

Available documents fail to confirm the story, nor do they quite dismiss it. They say nothing about Gray's venturing so far upstream, but they leave the possibility open. In fact, the log of *Columbia*'s second voyage was lost, but the relevant portion (containing the "discovery" of the Columbia River) was copied and saved in Boston. It states only this, for the one day the *Columbia* lay anchored off Gray's Bay: "May 15th. Light airs and pleasant weather; many natives from different tribes came alongside. At ten A.M., unmoored and dropped down with the tide to a better anchoring place; smiths and tradesmen constantly employed. In the afternoon, Captain Gray and Mr. Hoskins, in the jolly-boat, went on shore to take a short view of the country."

A second, complete log of the voyage, kept by a sixteen-year-old fifth mate named John Boit, refers to a landing also: "I landed abrest the ship with Capt. Gray to view country . . . found much clear ground, fit for cultivation, with the woods mostly clear from underbrush." Neither account records a five- or six-mile paddle upstream, so it remains conjectural whether the good Captain Gray ever graced our presumptive village with his presence after all.

We'll probably never know for sure. But he did give us his name, and a good name it is. Some may find "Gray" a bland, or even off-putting, name with melancholic or drab connotations. Not I. I find it a soft, succoring word, pleasing to the eye and ear. It accords beautifully with the skies, waters, and mists of the region, and so has descriptive as well as historic connections. The estimable navigator even used my preferred spelling.

The Board of Geographic Names decided some time ago to omit apostrophes. The resulting "Grays River" makes no sense. "Gray River" might be best as a commemorative, since a person cannot in any case own a river; but it is less euphonious. Thus, when confronted with geographical patronyms, I stick to the classical possessive. I trust my editors will allow me this tic.

Robert Gray's name decorates a number of other landforms and localities in the region, from Gray's Point to Gray's Harbor. In the case of Grayland, a coastal cranberry-growing village, who can say whether the name reflects the man or the prevailing hue of the days? I doubt that Robert relates at all to the prettiest of the Gray names, a hamlet still on the map called Gray Gables. I suspect this name came as much from the color of weathered cedar shakes as from some summer person's love of alliteration.

The only problem with these graynames arises with Gray's Harbor. Several days before she dared and beat the big river's bar, the *Columbia* crossed a lesser bar and became the first modern craft to enter a large bay at the mouth of the Chehalis River. Here the men traded copper and nails for furs, and John Boit marveled at the nakedness of the Indians. When they left, the sailors insisted over their captain's objections that the big bay be called Gray's Harbor; and thus it has been ever since.

The problem is that Gray's Harbor is much better known than Gray's River; and although there is no post office by the name of Gray's Harbor, mail intended for Gray's River often finds itself misdirected to Gray's Harbor and ends up in Aberdeen or Hoquiam. What is worse, I am often corrected about the name of my own town: "Where do you live?" "Gray's River." "Oh, you mean Gray's Harbor!"

But such ignominy as we residents of Gray's River must suffer should not be held against Robert Gray. After all, he wanted to name the feature after his employers—Bullfinch Bay would be a pleasant name, free of confusion. But I suppose Gray deserves his harbor as much as George Vancouver deserves his island and Peter Puget his Sound (which some think Gray discovered in any case). In fact, some hold that the Columbia River, Washington State, bears too much similarity to Washington, District of Columbia; and that we would be better off renaming the big river after Robert Gray.

More pertinent might be the fact that the Chinook Indians had been crossing the bar by canoe for centuries. If there were to be any other name for the Columbia, shouldn't it be Chinookan? Besides, if the Columbia were renamed Gray's River, what would we call this smaller stream that we have just navigated together?

Not that it would be a big issue. Gray's Harbor or no, Gray's River would still be mostly unknown to a great many people. There was a time when hundreds of Indian canoes ran with its advancing tide; a later time when steamers from Astoria called at the booming timber town, and the creamery cooperative served scores of dairies. Now the Indians and the boom are both gone. Only the occasional drift boat plies the Gray's anymore, in search of steelhead, or writers in their canoes, angling for words. Except for one day a year, when the river race runs, Gray's River flows quiet and mostly ignored.

So friends in Seattle will continue to correct me when I tell them where I live: "You mean Gray's Harbor, surely." And in interviews describing my home as overlooking the Gray's River valley, copy editors will continue to change the name of my town to one that doesn't even exist. That's all right with me. Robert Gray's river is worth knowing—but who needs a crowd?

In 1989, Washington State's centennial will take place. For their part, the townspeople of Gray's River are making a park by the river-shore and erecting a plaque to commemorate Captain Gray's supposed visit. Three years later, we will observe the bicentenary of that storied arrival. Perhaps some small celebration will be in order. But if no one comes, that's all right too. By all accounts a modest man, Robert Gray probably wouldn't mind. And as for the river, I suspect it would pay little heed either way.

# Part II

## Denizens

# Threads of the Green Cloth

Sage Hall, Yale School of Forestry and Environmental Studies, 1975. The empty chairs about the great round table began to fill as graduate students came into the room for their lunch break between classes. I lounged in a deep chair by the fireplace reading a *New Yorker*. Occasionally I joined in the conversations around the big central table, but today I just listened in.

"What do you think of that business about simplifying forests?" opened a young silviculturist-in-the-making.

"Makes some sense," replied another member of the ecology class that had just let out. "When you take a tropical forest of one thousand species, or even an eastern deciduous woodland such as we have here, and reduce it to a monoculture, it stands to reason you'd be lowering its stability."

"I guess so," the first replied. "Fortunately that doesn't apply to the timberlands out in the Northwest where I hope to work—they're so simple to begin with."

"Right," concurred the second. "Nothing but a bunch of conifers and a few alders—practically a monoculture to begin with."

A third joined in: "Yeah, that's right. I went to school in Corvallis and, man, are those woods simple compared to here! If you ask me, it makes sense to manage them for Douglas-fir. It grows great there, and there's nothing else to speak of."

I couldn't believe my ears as the talk went on. My own research dealt with the butterfly biogeography of Washington, and if it taught me anything, it was that the Northwest has its own brand of ecological complexity. Yet these novice foresters were describing the region as impoverished and extremely simple. I almost broke in to counter their misimpression, but then the subject changed to logging methods, and I had a lecture anyway. I figured they would find out for themselves, if they ever really got out into the Great North Woods and opened their eyes the way Yale was supposed to teach them to do. Over the past ten years, however, I've often thought of that incident and wondered what I would have said to change my classmates' minds about the "simple" woods of the West.

It is quite true that the tropics, and the eastern United States, support a degree of hardwood tree diversity much higher than here. I have been to a rainforest preserve in Costa Rica where well over one thousand species of woody trees and shrubs had been cataloged in one square mile. There were woods not far from Yale where fifty or a hundred species might be found. In contrast, the Willapa Hills support a mere dozen species of hardwood trees and not so many more shrubs. On those grounds alone, those students had something of a point.

However, I've often wanted to regress in time, shake them, and say, "Look! Hardwoods aren't everything! When you say we've got nothing but conifers, do you realize how many conifer species you're talking about?" Far from just Douglas-fir, Northwest woods harbor several true firs, two spruces, two hemlocks, four or five pines, larch, junipers, redwood, yew, and several cedars. Many of these grow in the Willapa Hills. When you add these to the several deciduous species, the diversity begins to look a little higher than those callow scholars supposed.

Nor should trees comprise the only, or even the best, measure of diversity and complexity (except perhaps to a forester). All groups of plants and animals should be weighed in the balance. If we do so, we find that certain of them mirror the hardwoods, and perhaps not coincidentally.

For example, a single two-square-kilometer preserve in Peru hosts over one thousand species of butterflies; the Willapa Hills have produced just over thirty. Our wood warblers number just seven or eight, whereas a single spring maple in the Connecticut woods might harbor thirty species plus. Higher botanical diversity certainly inspires herbivorous insect diversity, which in turn promotes insectivorous bird richness.

Yet look at mushrooms: according to the late Dr. Daniel Stuntz, my professor of mycology at the University of Washington, the Northwest forests are home to over three thousand species of large fruiting fungi—making them the world center of mushroom diversity in all likelihood.

What determines such figures? It is axiomatic in ecology that, as one proceeds north or south from the Equator, diversity of species tends to decrease as abundance of individuals within a species tends to increase. So in the tropical moist forest, there are many, many species of plants and animals packed in close together, with rather few individuals of most species; while in the Far North one finds few types of organisms, but the numbers of individuals (as with some mosquitoes) may be enormous. This trend has to do with the number of niches available in a hot, moist environment with even sunshine, as opposed to areas with greater extremes of cold and dry.

Naturally, different floras and faunas follow this law unevenly. Warblers, hardwoods, and butterflies adhere to it pretty strictly in the Pacific Northwest; mushrooms clearly do not. There are all sorts of temporal, ecological, geological, and historical reasons for this. Heavy recent glaciation, for example, reduced the natural diversity of Great Britain to an even greater extent than its latitude alone would have accomplished. The same may be said for Washington. Lying in the corner of the country certainly has suppressed colonization by species over time, making this region a biogeographical backwater, as it were.

Furthermore, species composition itself can affect richness in other groups. Conifers do well in the heavy rainfall regime of Willapa; yet few butterflies thrive in coniferous forests, nor do they colonize well across them. Such factors have not presented a barrier for fungi, which colonize by spore dispersal, above the treetops, often on the upper air currents, and find evergreen forests just dandy.

Clearly, ecological complexity may be arrived at and defined in many ways and affected by many factors. The question itself is complex. Simplicity, too, has many shades. But it is too facile, I believe, to say that we needn't be concerned with the diversity of the Northwest woods because they're just a bunch of conifers.

Complexity is not merely a matter of local pride. Another ecological rule, or at least a large body of evidence, suggests that the more complexity an ecosystem sustains, the more stable it is likely to be. This has importance for a later chapter concerning timber management. But it applies here too, for it raises the question whose answer rolled off that round table like a crystal ball—or rather, a bowling ball, for all the good it would do in securing the forest's future: if a forest, or other ecosystem, is to be adjudged *simple*, does it thereby forfeit its right to care, consideration of its qualities, and intelligent stewardship? I think not.

Complexity is relative; stability is absolute. Whether floristically diverse or less so, all landscapes face destabilizing influences if viewed as simple and treated accordingly. Of course in practice richness may make no practical difference as to treatment: the richest forests on earth are being felled and 'dozed as if they were Tinkertoys and Lincoln Logs, toy forests scattered across a concrete floor; whereas the "barren" tundra on the Arctic Ocean shores, as simple as an ecosystem gets by some measures, received at least some care and concern during the construction of the Alaskan Oil Pipeline. No system deserves less.

Now let's look at complexity in a different way. Number of species is not, after all, the only measure. There lives a subtle richness in the coniferous forest and its understory. If a hemlock radiates beauty in a green shade, how much more beauty flows from a million hemlocks? If a stand of old-growth cedar affects the spirit in an irreproducible manner, should that experience not figure in the complexity of the cedars' forest? And if primitive horsetails and tangled briers can help to hold the soil and build new forests from broken ones, who is to say they haven't value beyond the sum of their two species' tally?

The economic contribution of the flora comes not only in the form of pulpwood and two-by-fours, but also as ferns, berries, cones, mushrooms, and cascara bark, to say nothing of the watershed. But to many, the greatest treasure of the plants around us (beyond oxygen and foodstuffs) lies in their ability to arouse interest, awe, and admiration.

Perhaps nothing about plants touches our intellect hereabouts as much as growth itself. I can begin to empathize with my neighbors' penchant for cutting, spraying, slashing back the vegetation, opening up a parklike environment in their compounds, and constantly doing battle with plants when I see old pictures of how it used to be. This valley, whose broad green tablecloth I so admire, used to be a tangled, timbered floodplain—beautiful in a different way, no doubt more interesting biologically, but of no use to settlers. Ebba Sorenson's photographs, developed in what is now my coat closet, show in stages the reclamation of that bottom to the fields we see today. It was hard work, and only the restless jaws of cattle keep the scene the way it is today.

The growth of plants takes place with such incredible vigor in the rain world that one need but relax to see entire paths, fields, and buildings disappear from sight beneath a fresh green canopy. Of course, alien weeds have much to do with that; nothing in the native flora snatches land quite as fast as Himalayan blackberry. But the natives are no slouches, either.

As I write this on my front-porch table, the scouts of a gangly wisteria vine threaten to entwine my clipboard, encircle my very head, and snatch the pen from my hand. I pay little notice, then consider that the vine was cut to the ground (being a nonbloomer) not many months ago—and now it curls around the pillars and tugs on the eaves of the porch once more. Charles Darwin was so taken by the power of movement in plants that he wrote a lengthy book by that name. It contains the results of his many elegant and ingenious experiments designed to elucidate how plants move the way they do. These botanical studies by a geologist noted for his zoology combine the force of genius with the energy of unbounded nature to produce a monograph that holds up even today. As always, the clarity of Darwin's prose equaled his erudition. Though undecorated, straightforward, and rather dry, the great naturalist's words on the subject give one a clear sense of the wonder he felt when confronted by plants in motion.

When I wonder at plant power, I think of the marah. Also known as wild cucumber or manroot, this native cucurbit climbs high into the forests from its base beginning underneath. Marah, honeysuckle, and wild clematis, or old-man's beard, make our only true lianas—ground-rooted aerial plants with high-climbing vines. Marah dies back each year, then, beginning anew at ground level, clambers and twines fifty or seventy-five feet into the air each summer, bedecking hemlock, cherry, and bare slope alike with its grapelike leaves, watch-spring tendrils, and fruits like little green hedgehogs. The name manroot grows out of its enormous tuber. A new path down to the beach at Oregon's Ecola State Park recently revealed such a root in the trailcut. When we show it off, onlookers insist it must be some great stone imbedded in the soil. There lies the root of true growth.

Next to the speed and power of Northwest plants' growth, we marvel at their very prolificacy. As in the tropics, bare earth never stays bare for long here, except where erosion and leaching have created laterites too packed or poor for even the toughest colonists to take root. Even there greening will occur, but it must begin (as on stone) with mosses, and it takes forever. Otherwise, the greenwood grows all around, all around, and the earth goes verdant with the stain. The chlorophyll of a billion leaves leaks out over the land, lending it new life and giving the green light for growth.

In the rain forest of the Pacific Northwest, growth is nowhere more prolific than among the epiphytes. These are the mosses, lichens, liverworts, clubmosses, and ferns, chiefly, that deck the conifers, maples, cottonwoods, and alders, standing and fallen, as though some crazy upholsterers had run through the forest with their stores and stapleguns.

My mentor, Grant W. Sharpe, professor in the University of Washington's College of Forest Resources, performed (and that is the word) his doctoral thesis on the ecology of the temperate rain forest in Olympic National Park. As both a graduate student and a summer ranger for the National Park Service, he was admirably placed for such a detailed, painstaking study. The great westside valleys of the park—of the Queets, Quinalt, Bogachiel, and Hoh rivers—had never been investigated in depth. Grant decided to make the attempt, and he made the necessary commitment to extended trips into the wilderness over and over again, on foot, carrying seventy-five pounds of equipment, instruments, and supplies on his back. The result, in my view, well warranted the effort.

As Sharpe wrote in the preface to his dissertation, "Every American national park preserves some fundamental feature which distinguishes it from other national parks. . . . Olympic's unique feature is its westside forest—a rain forest. . . . The striking feature of these wet forests, in addition to the tree sizes [largest recorded individuals of four species of trees] is the luxuriantly developed epiphytic vegetation found upholstering the trunks and crowns of all trees."

Sharpe found that the crown habitat alone supported seventy-seven species of epiphytes, and a good many more joined the tally when trunk, base, log, and soil habitats were considered—a hundred and more species of epiphytes in the four valleys, below one thousand feet elevation. He went on to examine the ecological roles and impacts of these plants. As one would guess from their conspicuousness and sheer abundance, epiphytes play major parts in the forest's economy—just how much so, Sharpe could not know at the time. He concluded that understanding, protecting, and interpreting these little-known plants is essential to the future of the park.

This then is the apotheosis of moss, the elevation of lichen and liverwort, the glory of fern: a great national park, defined not so much by its glaciers, high peaks, Roosevelt elk herd, or roadless coastline, as by its forest and the lowly epiphytic plants that set it apart from all other forests. And while the diversity and integrity of the systems in Willapa are lesser by comparison due to lower rainfall (120 inches per annum as against 160+) and exploitation; and while Professor Sharpe would have found it impossible in Willapa to locate his 176 ten-thousand-square-foot plots of pure forest types for his sampling, so hacked about are the forests here; nonetheless in general, in aspect and impression, the epiphytes of the Willapa, the Nemah, the Naselle, and the Gray's display much the same luxuriance and exuberance as he found in the Queets, the Hoh, the Quinalt, and the Bogachiel river valleys.

When Grant Sharpe went packing up those Olympics rivers in the early fifties, it was believed (as he wrote) that "cryptogamic epiphytes, dependent on their supporting species for support only, have little effect on them." These aerial mosses, lichens, liverworts, and ferns "have no connection with the soil and are dependent on precipitation for their water supply. Nutrients are derived from rainwater, windborne particles, and decaying bark of the plant on which they grow." But "as they require no organic food from their supporting species, they are not parasites." In the context of the times, it made sense to conclude that, in all their luxuriance, epiphytes were opportunists that had no impact on their hosts other than occasionally increasing their susceptibility to windthrow.

That view has recently changed, and radically so. When I was in Papua New Guinea looking at butterfly farming for the PNG Wildlife Division, I met a young biologist at the Wau Ecology Institute. Excited rather than overwhelmed by the extravagance of the New Guinean rain forest, she told me that she was eager to return to the temperate rain forest to have a closer look at it. She too was unhappy with a view that held that a cooler, less species-rich rain forest was necessarily boringly simple; yet she was lured by the relative ease of working on a rain forest close to her home base of the University of Washington, where mere species identification need not occupy years of effort as it could in the tropics. (Not that identifying the mosses and liverworts and lichens of the Northwest was all that easy, as Grant Sharpe had found; but thanks to his work, much of that baseline had been drawn already.)

So in a nice piece of historical symmetry, Nalini Nadkarni began to study some of the same Olympic rainforests that Grant Sharpe had assayed three decades before. Yet, while Sharpe was restricted to fallen branches and wind-thrown trees for his glimpses of the canopy flora, Nalini employed modern alpine climbing techniques and equipment in order to penetrate the canopy in depth. As she carried out what some detractors called her Tarzan act, Nalini contemplated the great masses of mosses and other epiphytes. She found they could total more than one hundred pounds on a single bigleaf maple, or six thousand pounds per acre. And she began to question whether the relationship between epiphytes and their host trees was nutritionally neutral after all.

Stripping mosses from maples to be weighed, she made a remarkable discovery: there were roots among the epiphytes and their mats of detritus—tree roots. Never before noticed in the literature, these regular roots ran from branch nodes, often for tens of feet. They clearly existed to take advantage of the extensive nutrients captured in the moss and lichen mats, before they could be leached by the rain to the forest floor.

Suspecting something general, Nalini sought aerial roots among host trees in the richest rain forest canopy of all, in Costa Rica, and also in New Zealand's temperate rain forest and back in New Guinea, where I had met her. And in each of these places, she found a number of unrelated tree species taking part in the phenomenon of aerial rooting among the epiphytic gardens. Forget Jane; Ms. Tarzan was vindicated with bells on.

As reported in the February 1985 issue of *Natural History*, Nalini Nadkarni's findings greatly expanded our view of epiphytes. From ecologically beside the point and merely commensal with their hosts—providing neither harm nor help—they have grown in our understanding to describe a continuum, all the way from a few true parasites (such as mistletoe) through commensalism to truly mutualistic organisms.

As mutualists, epiphytes provide a useful link in nutrient cycling to those hosts able to take advantage of them while taking support and a place in the sun for themselves. Admittedly, the conifers can't do it: maples, alders, and cottonwoods can, in our woods, and no doubt their vigor is the greater for it. And the story of plants that grow on others, clothing the forest in thick green puffery and plumes, grows the more interesting for it. As Nalini put it, "Greatly excited, I realized that here was a shortcut in the rain forest's nutrient cycling system. Host trees were capable of tapping the arboreal cupboards in their own crowns." That *is* exciting.

Few of us will make discoveries of that magnitude as we unravel the threads of the green fabric. But in the plant world, there is no end of discovery to be made. Another botanist of my acquaintance, Cathy Maxwell, has found this as she has plunged into the task of compiling a flora of the Willapa Hills. Largely neglected by other plant prowlers, these scruffy hills offer an alpha-level of discovery almost daily for the observant and diligent explorer. Accompanied by her fisheries-biologist husband, Ed, and their Akita dog named Hyla, Cathy scrambles across the cutover ridges and reaches the peaks of Gray's River Divide. They explore wet creases in the land that might once have been bogs and clamber over stony balds where lilies and saxifrages find refuge from the loggers' slash heaps. A volunteer for the Washington Natural Heritage Program, Cathy makes her finds available to that bureau's rare-and-sensitive plant survey. Her lists have included many surprises for seasoned botanists who spend little thought and less time on the Willapa Hills, consigning them to the dustbin of used-up habitats: species new to the state, range extensions southward from the Olympics or northward from the Oregon Coast Range, and supposed rarities occurring in abundance.

These discoveries have been exciting to eavesdrop upon, but for our purposes Cathy's greatest contribution has been her Flora: a working list of all higher-plant species heretofore recorded in the Willapa Hills (which, for Cathy's purposes, are defined as an area somewhat smaller than I have described, for workability). The list gives us the basis for learning, studying, comparing, and otherwise confronting the plants we come across. Field guides and regional floras are fine for their purposes, but the latter are too inclusive and the former too exclusive to be immediately helpful in the local context. Cathy uses C. Leo Hitchcock's great tome, *The Vascular Plants of the Pacific Northwest*, as well as her own herbarium and that of the University of Washington, for her identifications.

Cathy faces a complex task: to define, for the first time, out of all the possible Northwest species, just which plants have reached these poor hills and survived all challenges, from superabundant rainfall to near-total disturbance by logging; and, to sort them out from the many possible aliens (no small task in a largely agricultural region abutting an active shipping lane and an interstate freeway).

By sorting out the possibilities into her Flora, Cathy saves us much of the effort of narrowing down the choices. That gives us more time to enjoy the plants themselves. And of course it alerts us to special things we might seek out or watch for, as well as potential host plants for butterflies I hope to find.

A quick glance through the Flora will serve to highlight the threads of this green cloth we call Willapa. Cathy has arranged the families alphabetically, which is fine for a survey since one would no more find plants arranged taxonomically in nature than by ABC. As I glance through *A Partial Checklist of Vascular Plants of the Willapa Hills and Lower Columbia River Area of Southwest Washington*, the names bring to mind images of particular stands, special inflorescences, individual trees, or seasonal spectacles—all panels and pleats in the curtain of the forest fabric.

**Aceraceae:** The very first name on the list, *Acer circinatum*, the vine maple, evokes clear pictures of arcs and hoops of this maple, which is a shrub or a tree but certainly not a vine, conferring a graceful concavity on the third dimension of the forest. So densely hang the mantles of the mosses *Isothecium* and *Rhytidiadelphus* on the vine maples that its own limbs may be lost to sight throughout the maple brake. Vine-maple leaves reflect a soft green light onto their own branches, which may be green too, especially in the rain. This green luminosity defeats the claim of total gloom upon the deep-wood. In the fall, it is the many-lobed paws of the vine-maple leaves that we catch red-handed in the act of painting the town. The other main maple is the bigleaf, and it goes gold, and between the two they lend the autumn most of its color. Club mosses and licorice ferns depend luxuriously from bigleaf maples, so that their boughs resemble the maned necks of great bull elk in rut, or camel blankets frayed from too many caravans. Such festooned maples led Grant Sharpe to name a particular spot in the Hoh River valley, the Hall of Mosses.

**Alismataceae:** Wapato, tuber-food of the Indians, its delicate white flowers against big, green arrowhead leaves appearing at water's edge.

**Anacardiaceae:** Poison oak; only in the driest parts of the hills, where the south sun strikes the walls of the lower Columbia Gorge. Heavy rain strikes it out of the scene.

**Apocynaceae:** Spreading dogbane, beloved of butterflies and bees.

**Aquifoliaceae:** English holly, come here with the immigrants, popping up prickly where never expected in the heavy forest.

**Araceae:** Skunk cabbage arises from the dead well before every Easter and continues to color the damp lowlands all spring, with its canary candles of spathe and spadix. By late summer, its leaves have grown to gigantic proportions, flopping like great green diapers across the drying swamps.

**Araliaceae:** A relative of ginseng, devil's club has no such life-giving properties; life-arresting might be more like it, if you were to fall on it. Its name, *Oplopanax horridum*, says it all. Tall wands lined with points like a Catherine wheel forged straight; broad, elephantine leaves similarly armed, red berries in fall that might have been distilled from its victims' blood.

**Aristolochiaceae:** *Asarum caudatum*, wild ginger, low and shiny with its green-heart leaves and witchy maroon flower the color of spoiled liver. Our only member of the great tribe of plants that pasture the caterpillars of birdwing butterflies and their relatives all over the tropical world. None here to eat it.

**Balsaminaceae:** Orange balsam or jewelweed, popping its seeds in the ditches and marshes after brightening their late summer.

**Berberidaceae:** Three of our common, graceful ground covers—vanilla leaf, Oregon grape, inside-out flower—arresting elements of nature's great herbaceous border and bandages to the wounded land.

**Betulaceae:** Our conspicuous birches arrived with the Scandinavian settlers to confuse the botanists, and the native red alders often become so heavily encrusted with white lichens that they are taken for birches even by the locals. Alder, the unsung hardwood mainmast of the Northwest woods, sprayed as a weed by the loggers, praised as nitrogen fixer by biologists. Even on its branches, beneath mats of moss, may its nitrogen-fixing, bacteria-bearing nodules be found, as Nalini discovered. And hazel, with its fuzzy leaves, scarcely bearing any filberts and what there are usually taken by the squirrels.

**Boraginaceae:** Bluebells and forget-me-nots, hound's-tongues and comfrey—little spots of blue among the general green.

**Callitrichaceae:** Water starworts lighting the surface of the dark pools.

**Campanulaceae:** More bluebells, properly harebells, simplest flowers in form and soft, pure hue.

**Caprifoliaceae:** A varied family and important, comprising the elderberries (red and blue), the white-globed snowberry, and the odd twinberry; the brilliant orange honeysuckle, punctuator of summer-forest drab; and Linnaeus' favorite flower, the delicate *Linnaea borealis* or twinflower—reminder of Willapa's boreal past and Swedish connections.

**Caryophyllaceae:** Pinks and chickweeds, subtle little flowers except for the introduced rose campion, raising bright flags beside old homesteads.

**Celastraceae:** The mountain box, tough little shrub that keeps to itself in the hills.

**Ceratophyllaceae:** Hornwort, not someone I know.

**Chenopodaceae:** Salt-tolerant shore plants so vital to the black brant in passage, the pickleweed (*Salicornia*); and fat hen, apparently also favored by fowl.

**Compositae:** Our largest family, home of asters and sunflowers and a host of DYC's: damn yellow composites. Yarrow and pearly everlasting, ubiquitous white-flowered weeds, color the waste places snowy and give nectar where none other may be found. Suksdorff's sagebrush blows in the bitter winds on bayside headlands. Groundsels and goldenrods make sunshine incarnate across the summer landscape, and asters daub the autumn salt marshes mauve. Thistles and ragwort cause farmers to cuss, dandelions and cat's ear plague the grower of lawns, while nectaring insects flock to these alien composite windfalls.

**Convolvulaceae:** Two introduced species, lady's nightcap and field morning glory, bind the soil as they defy the rototiller.

**Cornaceae:** Pacific and bunchberry dogwoods, uncommon here, spatter respectively the forest fringe and floor with cream; creek dogwood, spotted all along the sloughs and rivers, renders them carmine with the first frost. Its white flowers feed the spring azures, its white berries the wax-wings in winter.

**Crassulaceae:** Yellow stonecrops clinging to the basalt bluffs, furnishing forage for the compact larvae of the early elfin butterfly.

**Cruciferae:** The mighty mustards, including an arrangement of cresses besides the delicate pink *Cardamine pulcherrima* that nectars and forages the veined whites as the woodland begins to leaf out in spring.

**Cucurbitaceae:** Containing the aforementioned *Marah oreganus*, the manroot or bigroot.

**Cupressaceae:** The yellow-green fans of western red-cedar foliage all but hide its rusty, ropy bark from view. Most of the big cedars are long gone for shakes or fence posts, but here and there the canopy goes frondose, revealing an old arborvitae among the firs and hemlocks; or a range of ferny foliage shows where new ones have sprouted among the commoner conifers.

**Cuscutaceae:** The strange dodder, creeping parasite of the shore, sucking green from other plants and giving back a venomous orange.

**Cyperaceae:** "Sedges," they say, "have edges. . . ." They fill in the low, poorly drained patches between the trees, or bolster the boggy spots with bulrushes, which lack the prescribed corners.

**Dipsacaceae:** Alien teasel, pale purple flowers filling in the spiny head in stages.

**Droseraceae:** The insectivorous sundew of the bogs.

**Equisetaceae:** Horsetails, expanding out of road cuts like green bottlebrushes; rising high, then drooping with the heat, finally lying bleached and bony over the bare winter banks.

**Ericaceae:** No heather to take over from these vast clear-felled forests as there is in Scotland; but a bit of red madrona around the edges, patches of pipsissewa and kinnikinnick, and endless acreage of glossy-leaved salal. Impenetrable, salal gives the impression of one endless, interconnected plant, up and down the Northwest coast. Huckleberries of five separate colors, habits, and tastes; a plant for gnomes, a laurel, a Labrador tea; and a welter of wintergreens, lying low on the forest floor, taking no credit.

**Euphorbiaceae:** Moleplant and petty spurge, pretty plants brought to purge the land of moles.

**Fagaceae:** Oregon oak barely brushes the edges of Willapa, liking things drier.

**Fumariaceae:** Corydalis and bleeding heart—pink pronouncements on the state of spring, lacy statements in the mouths of ravines, fodder to the discriminating larvae of parnassians.

**Gentianaceae:** No blue gentians here, sad to say, just the introduced pink European centaury.

**Geraniaceae:** All of these introduced too, including an exotic pink geranium named herb Robert that likes overgrazed fields.

**Gramineae:** Grasses. What can I say about grasses, besides damn their pollen anyway in June, and hail to Cathy for dealing with them? Bromes and barleys, ryes and ripgut; fescues and foxtails, hairgrass and bluegrass—the names are as colorful as the plants and a good deal less troublesome to decipher. Of course, many of the grasses are introduced, like the invader *Spartina alterniflora* on Willapa Bay, and grassland is much more prevalent since the advent of agriculture. But here, as anywhere, native grasses form the batting for the landscape, holding it all together and claiming the bare soil as soon as it's ready for roots.

**Grossulariaceae:** Currant events, the most newsworthy being the first red currant to burst the bonds of winter buds on the cliffs above the river. Stink currant spots the damp swales with cerulean blue, coming from a flower you never noticed.

**Haloragacaeae:** South American water milfoil, waterway hog, inspires panic in officials when it appears and should do the same for native plants.

**Hydrangaceae:** Idaho's state flower, mock orange, just grazes the sunnier southeastern corner of the range, where its fragrance strokes the breeze like silk.

**Hydrocharitaceae:** It is no wonder that waterweeds of the frog's-bit family should be present in this watery land.

**Hydrophyllaceae:** Ditto for the waterleaf family. The Sitka mist-maiden must be mentioned for her lovely name, though I haven't yet met her in her foggy haunts.

**Hypericaceae:** We have two understated Saint-John's-worts, but the common one is introduced: as with so many other alien weeds, it found disturbed conditions here perfect for its spreading needs and was preadapted for them by centuries of occupation alongside human settlements in the Old World.

**Iridaceae:** Irises, represented by blue- and yellow-eyed grasses. Europe's yellow flag, and the blue flag of Oregon, *Iris tenax*. The Oregon iris leaves a soft, surprising violet note on the roadside, a figment of before on a clearcut ridge.

**Juncaceae:** ". . . and rushes are round." Of eighteen rushes present, seventeen are indigenous. *Luzula*, the woodrushes, beautifully named both in English and Latin.

**Juncaginaceae:** Quillwort, arrowgrass, presumably pointy plants both.

**Labiatae:** Mints, packets of freshness whether crushed by boot, fingers, or teeth. Self-heal, a blue accent to the mowed verge and welcome nectar bank; mad-dog skullcap, whose acquaintance I must make; Yerba Buena and Mexican hedge-nettle, lending a Latin note to a land that seems anything but.

**Leguminosae:** The great sisterhood of peas—*Lathyrus*, *Lotu*s, lupine and vetch, and all the clovers as well—thirty species of colorful confusion, serving the dual purpose of enriching the soil with nitrogen and confounding the naturalist. The close dependence of animals on legumes reflects in names like deervetch and buckbean; others appear in descriptive names such as bird's-foot trefoil, hare's foot, and tomcat clover. Aliens loom large: everlasting peavine and common vetch paint verges pink and purple; Scots broom, gorse, and trefoil burst into yellow bloom high and low and may be welcome for their color if not for their bad habits and tenacious growth. Native giant vetch feeds tailed blue larvae.

**Lemnaceae:** Shy *Lemna minor*, the water lentil.

**Liliaceae:** Still more bluebells, the English ones; and onions and fritillaries and Solomon's seal. Pink fawn lily, scarce along wild riverbanks and in other secret places. And white trillium—most prominent of all the modest spring array of wildflowers in the Washington woods; nothing modest about it—three bright white spear-petals that go rose with age, above three green spades of leaves that last all season.

**Lycopodiaceae:** Ground pine's chartreuse chains, writhing slowly across bare mineral soil.

**Lythraceae:** European purple loosestrife, giving color that never was to latterday shorelines, while taking over wetlands.

**Malvaceae:** Muskmallows and checkermallows, hosts of the West Coast lady butterfly.

**Nymphaceae:** Water lilies, aquatic extravagances, shelter to the muskrats, roadblocks to canoes, namesake of the valley Lily Pond.

**Oleaceae:** The elegant Oregon ash gives itself away with unique greens and golds, where it grows along the lower waterways.

**Onagraceae:** Never as prolific as in the Yukon, neither absent, fireweed arises orchid-purple each high summer in waste places. I prefer the English name, rosebay willowherb. Enchanter's nightshade.

Evening primrose, night-enchanter of moths with nocturnal fluorescence and perfumes.

**Ophioglossaceae:** Not a fern, the leathery grape-fern belongs to the adder's-tongue family.

**Orchidaceae:** Not many orchids here, though it is the largest family of vascular plants in the world. Yet the tall white bog orchid stands along valley riverbanks like pillars among lotus and sedge, and the spotted coral root surprises dank places on the forest floor. Tway-blades, ladies' tresses, rattlesnake plantain, and helleborines add to the orchid flora of Willapa.

**Oxalidaceae:** Oregon oxalis—wood sorrel—call it shamrock. Strike through the timber to a once-burnt ridge, where tall-again hemlock shades the humus rug and nurse logs serve as floorboards. Here, you will see acres of shamrocks spreading their triptychs of solid green, one by one, in a tapestry of incredible complexity and simplicity together. They blossom briefly white and pink, then decay into winter to help build the true sod. Nibble the lemony foliage, delight in the limitless symmetry of the sward, before looking up at last to regard the evergreens above.

**Pinaceae:** Here are the big parts, the seven stars, the conifers that are supposed to be so simple. Best just name them, their portraits (from various angles) generally appear elsewhere anyway: Pacific silver fir, grand fir, noble fir; Sitka spruce; lodgepole pine; Douglas-fir; western hemlock. These seven (along with the western red cedar) are the *raison d'être* for many of us here, in one way or another; they form the forest's flesh and caused the forest's fall. With cedar and yew, the pine-family plants comprise the framework of the story, if not the density, of the plot of the plants of the hills. And though much altered from their original grandeur and mass, the evergreens still dominate nearly every scene, and make the place the wintergreen land it is.

**Plantaginaceae:** Two exotic plantains inland—tough rosettes surviving where log trucks and lawn mowers roll right over them—and one indigenous species on the shore.

**Polygonaceae:** No buckwheat occurs in these hills, automatically excluding a number of butterflies; but the abundant and lovely purplish coppers thrive on several of the knotweeds and docks in this family. Japanese knotweed forms canebrakes where it has taken hold, such as Columbia River dredge spoils heaped on Puget Island, with big green leaves that turn field corners gold in fall.

**Polypodaceae:** Ferns: maidenhair and lady, deer and licorice, sword and bracken, others. I came for the ferns, in part, and then proceeded to learn almost nothing about them. Ferns have a way of making most other plants look coarse. Most notable: licorice ferns hanging like so many silkmoth antennae from broad laps of maple, or decorating a cliff at Skamokawa; leather ferns on old-growth spruce by the sea; sword ferns, head-high, climbing out of the crotches of the land; and maidenhair, fine black pelage curling out of the damp recesses of ravines.

**Portulacaceae:** A spring roadside without candyflower? As rare as a forest ramble without a nibble of miner's lettuce.

**Potamogetonaceae:** Waterweeds called pondweeds.

**Primulaceae:** No primroses in the May woods here, but western starflower instead.

**Ranunculaceae:** Baneberry, bugbane, and buttercups; creeping buttercup, an international terrorist, steals pastures and gardens wholesale—but does so beautifully. Spurred beauties in the canyons, red columbines, blue larkspurs. Yellow marsh marigolds on the salt marsh, white on the high divide.

**Rhamnaceae:** Cascara belongs here—small tree of gracefully cut ovoid leaf, deeply veined, going bright red in October; more common in the Willapa Hills than anywhere else I know, even to the point of forming pure stands, in spite of the commercial bark-strippers.

**Rosaceae:** So many roses by so many names, it's hard to choose. Must mention Indian plum, first and freshest green leaves to appear against March's backdrop of needletrees. Bittercherry and crab apple, white flowers in the young year's hair. Nootka rose, a fine species that might have been the progenitor of all garden roses, had rose culture been in the repertory of Northwest Coast Indians. And the gamut of *Rubus* species that fill our berry pots from midsummer on: blackberries, dewberries, thimbleberries, salmonberries, and others. If hemlock forms the weft of the woods, then salmonberry makes the woof—at least in countless swales and slopes where the yellow canes run at all angles to the walker who would needle his way through. Often the appearance of a magenta rose among the salmonberry provides the first outrageous sign of spring's forthcoming. Thimbleberry waits many weeks to put out its floppy, big white hats of flowers, direct contradictions of its specific name, *parviflora*. Goats-beard, queen of the forest, strawberry, mountain ash, and ocean spray each make their pale appearances in the proper places and times. Douglas spiraea, most ethereal of blooms, softens the marsh edge, soft, pink, fragrant, scarcely fitting its alternate name of "hardhack."

**Rubiaceae:** The sticky, scented bedstraws; and sweet woodruff, established on homesteads like ours.

**Salicaceae:** Only half-a-dozen willows to sort out here, and someday I will. Black cottonwood, sentry, looking sentient along the valley floors, where a few were spared for cattle sunshades and umbrellas.

**Saxifragaceae:** A wonderful array of pretty flowers, spurting from cliffs, tufted among the forest understory, standing forth from streamside lawnlets. We each have a favorite: alumroot or mitreroot, fringecup or foamflower, golden carpet coloring the April streamside; mine is the irresistible piggyback flower, named for two great botanists: *Tolmiea menziesii*. Rocky seeps running with meltwater or stored rainwater in the spring seldom fail to toss white sprays of saxifrages among their water drops, more delicate than any florist could conjure.

**Scrophulariaceae:** The most prominent beardtongue in the district is no native, but the flashy, extravagant foxglove. These magenta trumpets cover whole hillsides recently bereft of their timber, so that their introduction might be forgiven or even welcomed. Also, a rainbow array of elaborate natives: in that same cool spray of spring seeps, monkeyflower spreads its yellow faces and superlush salad-green leaves that vary dramatically in size. Cerise and indigo penstemons seek instead the hot rock walls. Paintbrush may bloom in any month, while snowqueen blossoms but briefly in March. Brooklimes and speedwells, genus *Veronica*, fleck the grassy places with blue nectar-pots for the spring azures.

**Selaginaceae:** Club mosses weave entire sections of Nalini's tree-trunk mats, swinging from maple boughs. They resemble moss, but are closer to ferns.

**Solanaceae:** Two attractive aliens, bittersweet and black nightshade, in the potato family.

**Sparganiaceae:** Bur reeds—tall green spikes that prick the surface of lowlying wet places.

**Taxaceae:** Woodsmen once watched out for yews on behalf of Puget Island boatbuilders, who prized the qualities of their hard, red wood; modern loggers consigned them to the slash heap, until its constituent taxol was discovered to be a powerful anti-cancer agent.

**Typhaceae:** Cattails, known in England as reed-mace, but everywhere *Typha*; you can forage for the pollen or the tuber starch, or simply the feeling of stripping the winter fur from the brown tail and sending its seeds to the wind.

**Umbelliferae:** The carrot family, in all its diversity. Angelica and water hemlock and cow parsnip and lomatium all nurture the anise swallowtail in its larval season. Sweet cicely and water parsley lend the ditches their feathery greenery and the flora their comely names. Queen Anne's lace was imported seemingly to line every country road.

**Urticaceae:** Stinging nettles, of course; a pain, yet host to red admirals, satyr anglewings, Milbert's tortoiseshells, and many other insects.

**Valerianaceae:** Plectritis and valerians describing a palette's worth of pinks on the rocky bluffs in spring.

**Violaceae:** Our violets are chiefly yellow (pioneer and evergreen), though the marsh violet is a clear violet-blue. Spring azures nectar on the lemony flowers of *Viola sempervirens*, and hydaspe fritillaries lay their eggs on *Viola glabella*'s leaves.

**Zosteraceae:** Eelgrass, substance of the salt marsh, city of teeming life at Willapa's door.

I hope this passage through the flora will have aroused your own personal portraits of the plants you know. I've mentioned just the least part of the list. Altogether, it totals some seven hundred species in eight-four families so far, and it grows each season through Cathy's diligent pursuit.

This, then, is a glimpse at the plant life of the Willapa Hills: the least complex part of the Northwest woods, described by my fellow students those years ago as "too simple to bother about." It may be a simple enough community in some respects: only one paintbrush and five lupines to worry about identifying, for example. And, to be sure, the number of tree species is far fewer than in Connecticut, let alone Costa Rica. Still, I hope I've shown that there is more to complexity than trees. Two or three thousand fungi; hundreds of mosses, lichens, and liverworts; all these and so many other fibers of the green fabric go to make up the tapestry of the Northwest woods. I'd like to see the quantitative ecologists erect a mathematical model to take it all into account. Simple, indeed.

Yet, such attitudes promote a state of artificial simplicity. They justify the creation of a forest monoculture, a regime that loses species and texture over time, as our flora doubtless already has. Yes, I've often wished I could return to that roundtable discussion, and join in. "Now wait a minute," I'd begin.

# Slug Love and Spider Hate

Something about snails drove me wild as a child. A youth of the plains, I had no access to the sea, nor to the moist haunts of land snails. Conchology and Colorado scarcely went together, but this failed to dim my passion. At least I could obtain seashells, from the Denver Museum gift shop, from dealers' catalogs, from my Shell-of-the-Month club, from a grandfather calling on aunts in Florida. So marine mollusks captured my passions in toto from age seven to eleven, and A. Hyatt Verrill's *Shell Collector's Handbook* became gospel as well as treasure map. The seven seas were mine in Aurora, Colorado, shining through the glossy nacre of my cowries and cones, olives and volutes, conches, helmets, and whelks.

Yet land snails held out a special mystique, and I don't doubt I would have traded toes to find some. When my father could be inveigled into stopping the car beside prairie creeks, at least I could gather tiny stream snails from the vegetation; unimpressive, but they were snails. Then when my grandfather took my brother, Tom, and me back East in the Packard and we called in at the Big Spring in Missouri's Ozarks, I had no eyes for the usual souvenirs: for there I found my first land snail, laying its silver streak on the wall of a cave beside the blue pool of the spring. I was transported with excitement, and the creature occupied me for the rest of the long, hot trip, then died.

I often wonder how I would have reacted to the ordinary abundance of varicolored helices in the Hawaiian rain forest, or even in an English garden, had I been emplaced there at, say, age eight. Any English garden would have been a magic garden then; a thrush's midden would have eclipsed a pirate's treasure trove in glory. Come to think of it, my other total passion in those days was for suits of armor—just the objects, divorced from their purpose; I never cared for warplay or toy soldiers. (What, the analytical reader might ask, does this say about my sense of confidence in childhood, that my objects of keenest covet were snails and chainmail? "Gimme Shelter," or what?) I reckon a trip to the Armory of the Tower of London, followed by a ramble in a rainy, snail-ridden churchyard, would have finished me off from overstimulus: cardiac arrest at eight.

Instead, I nearly perished from withheld gratification. *The Young Collector*, by Wheeler MacMillen (New York: Appleton & Co., 1928), informed me clearly that "land snails can be found almost anywhere," and that "slugs and snails abound in wooded places. . . . Look for them in ditches and river beds and muddy banks, for as a rule they love moisture." Oh, I looked—and looked and looked—in every moist and muddy place I could find, which were rather few in the Colorado summer, except when thunderstorms filled the ditches to overflowing and muddied the raw, young gardens. I combed such places, young slug me, in my slime track of desire. Thank goodness my mother was a naturalist. But it did no good.

Eventually, having far overspent my allowance and more on costly exotic mollusks, and with my land-snail collection still consisting of one, I decided on a shift of emphasis. One could hardly call me fickle; for four years I devoted myself to my beloved snails, but they spurned me all the while. So when I began to notice bluish skippers and black swallowtails in the parks and gardens, who could blame me for switching my affections? At last I fled the muddy banks for the sunny fields of butterflies, and satisfaction was mine.

Horror, too, as it turned out. Somehow, I had gathered about me a terror of spiders and their webs at least equal in intensity to my ecstasy over snails. Hanging about in the rain, in my prereformation days, I ran little risk of butting heads with spiders. But as I began to take to the sunny glades in pursuit of Psyche, out jumped Arachne and sat down beside me, and I freaked out.

I'll never forget my first face-to-face with a golden-and-black garden spider, a great *Argiope*. I simply *ran*, and screamed, and never looked back. And I wouldn't come home until my father arrived on the scene and guaranteed safe passage. All my brother had to do to excite revulsion in my heart (and get what he wanted) was to threaten me with an innocuous piece of cobweb. "Dust!" he exclaimed in mirth and amazement, "he's afraid of *dust*!" It always worked. Or he would lure me into a grandmother's cellar, then turn the lights out and proclaim the presence of spiders. I'd do anything to be saved.

Of all the monster movies, the black-widow scene in *The Incredible Shrinking Man* and the strange western *Tarantula* were by far the most terrifying to me, yielding many nightmares. "Here be spiders" might have been my watchword for panic, and by comparison I never could understand the literary and mythic concern with dragons. If I were to awaken and find a jumping spider on the ceiling over my bed, I'd pupate instantly beneath the covers, school or Saturday plans notwithstanding. I'd as soon perish as face it, because if life had to be like that, well, it just wasn't worth living.

Obviously, I couldn't go on this way. Fascinating cellars and caves and barns were psychologically out-of-bounds, not to mention many an otherwise seductive butterfly haunt. Running and screaming became socially penalizing as I approached adolescence, and I had one strike against me as a butterfly collector already. I tried to adopt an attitude of camaraderie toward spiders. After all, weren't we colleagues in the pursuit of insects, differing chiefly in the nature of our nets? Through the power of positive thought I came to be able to tolerate the knowledge that there were spiders in the same county as me, or even large field, but if I should come up against a bulbous-bodied orb weaver in the middle of its web, or inadvertently touch either animal or silk, especially with my face, the bottom fell out of my belly once again.

Several experiences helped me to overcome my arachnophobia. First, I worked as a summer ranger-naturalist in Sequoia National Park in 1969. One of my interpretive duties involved leading groups through Crystal Cave. Famous for an endemic spider, this fine marble cavern was protected from unauthorized entry by a wrought-iron grate in the form of a spiderweb, with a big black spider perched in the middle for a handle. But no one had seen the noted resident for years. So I dedicated my lunch hours to crawling about with a flashlight, in pursuit of the endemic troglobite. I found many of these especially unintimidating, little pink spiders, and watched them trap springtails in their tiny, weak webs. These became my first spider friends. (Furthermore, I encountered my first tarantula in Sequoia's chaparral that summer and while I did not pick it up, I did photograph it on my boot. I was coming along.)

Next, in 1971 and 1972, I dwelt in a 330-year-old, timber-framed and part-thatched yeoman farmer's cottage in England. We had no choice but to share our quarters with the immense English house spiders, *Tegenaria gigantea*. Since their webs were not orbed and their bodies not bulbous—the real psychological red flags—I could cope with them, despite their impressive size. By midwinter, unable to have a cat, we adopted the largest of the spiders as a pet, christened it Cat, and became watchful of its behavior. Perhaps I imagined our shared regard, but the remarkable actions I observed on Cat's part (following me from room to room, etc.) certainly raised my own regard for Cat and its kind. For the first time I felt real affection for a spider, and that was a great (if anthropomorphic) step forward.

In 1974 I was privileged to watch the great entomological ambassador of the American Museum of Natural History, Alice Gray, in action. She presented her live classroom demonstration for a meeting of the Xerces Society, including the laying on of tarantulas. I declined, but watched, rapt. Then a couple of years later in Colorado, when I had the chance to handle a friend's splendid Mexican red-kneed tarantula,

I took it. I've never forgotten the friend's name (Cosmo Blank) nor the lovely feeling of a furry great spider perambulating my palm.

Finally, my work and travels in Papua New Guinea in 1977 brought me inexorably into contact with a great array of tropical spiders. To my immense relief, none of them ever pounced from overhead, ate my face, or penetrated my sleeping quarters. And they were, of course, both beautiful and fascinating. I photographed them by the dozen and lost my fear of confronting them on the jungle trail (although, admittedly, I seldom went first; or if I did, my stout net served as a webwand, as it does today).

But my ultimate trial came one day when I was tramping the bush on Kiriwina, one of the Trobriand Islands, in search of eclectus parrots and Ulysses swallowtails. As ever, a group of curious villagers accompanied us. We encountered a very large, very bulbous, very leggy spider astride its three-foot web. I was about to snap its portrait, when, mantislike, a brown arm shot out, nabbed the treat, and popped it into a smiling child's mouth. Crunch, crunch. I reeled but withheld a retch as she popped the body like a grape under her tongue. The legs cracked like those of a prawn. My eyes must have been as big as hers had grown when I'd stepped from the plane, a bearded man with a butterfly net. Now here was the great white hunter turned green from witnessing a simple, everyday act of arachnophagy.

OK. Incident withstood . . . or so I thought. We had gone on a short distance when the snatch happened again. Only this time, the dainty was proffered to me, with many giggles. I won't drag this out. I declined the kind offer, and no one was offended (they never expected me to take it). But I did ask the girl to place the spider on my forearm, the whitest, most sensitive part. I felt every leg and watched its fangs and spinneret intently as the spider struggled for a hold. I photographed it, then tossed it gently into the bush before someone could eat it.

Sometimes I have regretted my lack of adventurousness in turning down the spider-snack. W. S. Bristowe, traveling in Siam a half century ago, resolved to sample the indigenous insect foods of the Laos. He found that soft-bodied spiders had "a nice crisp exterior and soft interior of soufflé consistency which is by no means unpleasant." The giant spider *Nephila*, an orb weaver, he found tasted like lettuce or raw potato; and he suffered no ill effects. My tale would be better had I possessed Bristowe's fortitude! But I suspect that, given the opportunity again, I would still pass. Anyway, I survived, and that was the transcendence of my unfortunate fear.

Just as well, too, for when I came to Washington State to live, I found it to be Spider City. In fact, one of my first jobs took me into an abandoned farm in rural Snohomish County, a site I named "The Vale of Spiders" for the thousands of alien orb weavers (*Araneus diadematus*) that slung their lines between every available pair of supports. I came to find that one cannot enjoy late summer in western Washington without a tolerance for these and the many native spiders, for they lie in wait everywhere for flies, grasshoppers, and arachnophobes. (My mother wanted to bring me to Seattle as a child. Had she succeeded, infantile paralysis would have been my certain fate. One look at the spiders outside and I never would have moved again.)

I've often been thankful that I dealt with my problem before I settled here for good. I still don't sleep with spiders by preference, though leggy *Pholcus phalangioides* dangles over my bed unhindered; and big chestnut-bodied spiders (not the same species as Cat, but the even more robust *Callobius severus*) hunt from many a cranny around the window sashes and commonly land in the tub. Nor do I gobble them down for snacks, unless they hide on the wrong side of a blackberry. But I am quite happy to live among them. And that is good, because live with them I do.

Last autumn, we invited Rod Crawford, Spider Man at the University of Washington's Burke Museum, to be our guest for a week as he surveyed our arachnofauna. Even hampered by an October rainier than usual, his findings were impressive: over fifty species of spiders in the house, grounds, and nearby habitats, and a large array of harvestmen, pseudoscorpions, isopods, centipedes, and millipedes as well. Rod feels the count could be greatly enhanced with better weather and a wider seasonal window of sampling. Swede Park, inside and out, year-round, is happy hunting grounds indeed for the likes of Rod.

Spiders, as one learns, enhance one's life even if one is not an arachnologist. Of course they consume flies and other bothersome insects in large numbers; but consider their beauties too—such as the crosslike pattern on the backs of the big female orb weavers so common hereabouts. Need so much as a word be written about the beauty of webs in the morning, in a misty place so generously hung about with them? Their short-lived lambency so perfectly reflects the nature of light and substance in a land where the distance from dead gray to brilliant silver is an angstrom or less, where form and formlessness mingle intermittently in the mist.

I grow more and more appreciative of spiders' ability to disperse when I watch the rain of gossamer in summer: exodus of spiderlings ballooning wherever the breeze should take them. Or when, in April, the yellow crab spiders always appear in the daffodils: from where? Most of all, spider behavior fascinates and instructs. The eternal spinning, consuming, respinning of the web, in so many species-specific patterns; the callidity of the hunt, and the specialized manner in which each species wraps its hapless prey in no-getaway silk. Then there is spider courtship.

Where I sit and as I write, I watch the attempted mating of a pair of cross spiders in a corner of the porch. The much larger female has now repelled the skinny, randy male (apparently taking him for prey—he has to be nimble) twice. She's sucking on a well-upholstered crane fly (atavistically familiar: they're both European species) and here he comes again; she can't be bothered. Climbing the sky (the web invisible against gray cloud) these silhouette spiders endlessly engage one another as the summers pass; only the individuals change. Just as endlessly may they enchant the watchful, he or she who notices and is not so silly as to turn away in fear or disgust.

Sun strikes the pair, giving them color; they engage again. In and out, he finds her line, the crane fly flexes its only free knee as the spiders pat pedipalps; then she strikes, he retreats, and you wonder how spiders ever do get mated. The male returns and supplicates with rapid strokes of his front legs, until finally getting the message she pulls in her own legs and goes still. The male addresses the female belly-to-belly, hops from his strand to hers, and they couple. A few seconds pass, and he falls abruptly away on his safety line. Four times this action repeats, as the crane fly waits, until the male retreats to his pillar. A few minutes later he comes in again on the web. She's ready this time, her predatory mode suppressed. Coupling occurs quickly, lasting three seconds.

Fascination with this process led me to investigate what is known of how it works. Rod Crawford, the arachnologist who performed our Swede Park spider survey, supplied the following explanation. First, the male recognizes the species and maturity of the female by the presence of her pheromone in the web. Courtship takes place through vibratory means, involving a species-specific pattern of plucking and shaking the web. In its last phase, courtship involves tactile and contact-pheromone signals. This complex process explains why it takes so long to prepare for a momentary mating. This lock-and-key series of stimuli and responses serves an important function, for accidental mating with the wrong species would mean wastage of genes. Therefore, elaborate barriers to hybridization (such as this courtship sequence) have evolved.

But it doesn't stop there. The male has previously deposited liquid semen on a thread or "sperm web," which he has drawn into his palps prior to initiating courtship. The sperms, still in a liquid form, pass to the female through the "business end" of the palp. Fertilization requires the "docking" of the palp in the structure of the female especially modified to receive it. This may be preceded by many minutes of fruitless attempts to engage the palp. When coitus finally occurs, it results in the detachable, sperm-bearing cap breaking off. So male spiders of the orbweaving genus *Araneus* can only mate twice since they have two palps. This is an unusual behavior: most male spiders can mate repeatedly, and it lasts much longer, up to an hour.

The female cross spider has two sperm reservoirs, one for each of the male's pedipalps. She may be mated only once on each side. This ensures that a successful male's genes will be passed on—an evolutionary reward that could help justify the expensive adaptive investment in such an elaborate mechanism for mating.

On a nearby web, two males approach a single nonreceptive female, and I begin to watch it all over again. How much better this is than running and screaming . . . and how much more parsimonious.

I couldn't live here unless I'd made my accommodation, southwest Washington being such good spider country. The irony is that it is also great snail country. At last, thirty years too late, I've landed in a place with snails right in my own backyard. No longer does *The Young Collector* fib to me—everywhere it says to seek, I seek and I find. Thirty years in the dimming, the joy at last achieved can't be as sharp as it might have been in the fruitless anticipation of youth. Still, something of the old gut thrill brought on by the mere suggestion of the coiled creature comes back, so that curiosity wakes wide and the finding satisfies in the end. "The spiral trail," is how Victor Scheffer, doyen of Northwest naturalists, refers to this unorthodox hunt in reporting his "notes of a snail watcher." The noted author of *Year of the Whale* and *Year of the Seal*, accustomed to bigger game, nonetheless finds the land snails of the Puget Sound Basin well worth his watching.

In the Willapa Hills, I watch much the same assemblage of species: the beautiful yellow-and-brown striped faithful snail *Monodenia fidelis*, similar to *Cepaea nemoralis* of English thrush middens; its small relative, *Vespericola columbiana*, whose shell is clothed with a fine pile of short fuzz; *Allogona townsendiana*, named along with a warbler and a chipmunk for the great naturalist J. K. Townsend, a big, chocolate snail with a chalky cast and taupe body; and the Vancouver green snail (*Haplotrema-vancouverense*) of milk-white foot and moss-green shell. Each of these exhibits its own habitat distribution and behavioral and ecological patterns as it forages across the wet, succulent greengrocery called by humans the Maritime Northwest.

I think of particular snails in particular places. Townsend's snails working the earth beneath stripped trunks of vine maples, where elk have rubbed themselves free of their itchy velvet. A faithful snail, massive, mauve-bodied, perched on the mossy crotch of an autumn-gilded maple, eight feet off the ground: another beside the westernmost Oregon iris I've ever seen in these hills, gracing together a clearcut bluff above Gray's Bay. Or a Vancouver green snail, browsing a chanterelle among hemlocks beside the town dump, showing uncommon good taste in choice of entree if not establishment. The species diversity may not be high, but the pleasurable surprise of coming upon any one of these tentacled gliders never seems to diminish with familiarity. And it is especially good to reflect on the dry days of my youth and know that, almost any damp day now, I can saunter out in the rain and find green snails—green snails!—abroad on the land.

One other species that can be found locally has not excited uniform enthusiasm. Called *Helix aspersa*, the European brown or garden snail, it is an introduced species that has become a substantial agricultural and garden pest in some areas. An inch across, quite conical, and banded with brown and beige, *H. aspersa* makes a striking sight that cannot readily be confused with any of the native species. While most people revile it as a competitor in the garden, it has earned a certain dedicated clientele. Since the true escargot (*H. pomatia*) has become uncommon in Europe (and indeed endangered in some districts) from overharvesting, the smaller garden snail has taken over as a quite serviceable substitute for gourmet escargot, both here and abroad.

My friend Tony Kischner, proprietor of the nationally noted Shoalwater Restaurant on the nearby Long Beach Peninsula, received a curious letter from the Washington State Department of Agriculture. Tony reports that the officials requested his cooperation in controlling the European brown garden snail, asking him to cease and desist rearing it for gastronomic purposes. The Shoalwater had long since given up the chancy practice of snail husbandry in favor of foraging for wild mollusks in the nearby woodlands, where they occur commonly. The results draw praise from many of the restaurant's customers, who probably neither know nor care that the French consider *Helix aspersa* a decidedly inferior snail.

Tony had the perfect answer for the agricultural officer. "Perhaps you should not overlook the 'French Solution' to snail control; this would require only that we educate our own people to consume the tasty, protein-rich morsels faster than they can consume our crops. . . . I would suggest that our practice of plucking mature snails from the local countryside to serve in our restaurants is not inconsistent with your Department's goal of reducing or eliminating the *Helix aspersa* population in the state. We just find that garlic butter is a much more palatable way to do away with them than any available pesticide."

This talk of edible snails brings to mind my first experience with the true escargot, *Helix pomatia*, the gastronome's gastropod. While rambling over the orchid-rich North Downs in Surrey with my good friend Jeremy Thomas during a springtime in Britain years ago, I spotted a gigantic live snail. Fully four inches in circumference, the spherical brown shell could quite properly be called awesome. Jeremy identified it as an edible snail, otherwise known as the apple snail or Roman snail. The former name, he explained, derives from the species' shape; the latter, from its presumed introduction to Britain by the Romans. Even during the days of the Roman Empire, this mollusk was appreciated for its succulence and brought along to lend its considerable weight to the support of the legions.

The snail in question turned out to be amorous as well as enormous. Many snails, of course, are hermaphroditic, but most species exchange sperms anyway in order to maintain genetic variation in the population. As I stroked the generous foot of the downland snail, it expanded and stretched to cover my entire index finger with its slick flesh. This is exactly how one snail behaves with another, but I didn't realize that and was caught all unawares. All of a sudden I felt a small, sharp pain as something pricked my finger. It was the love dart of the Roman snail!

I knew that snails exchanged love darts *in copulo*. However, I always thought they contained the sperms themselves. It turns out, as later reading has revealed, that sperm exchange is effected in the manner usual for "higher" animals. The calcareous darts precede ejaculation, and they probably serve a readiness role, enhancing the partners' receptivity.

Adrian Forsyth, in *A Natural History of Sex* (Scribners, 1986), points out that the love dart is composed of calcium carbonate, an essential material for snails. A hermaphroditic animal faces more of an investment by receiving sperms than by donating them, since it will have to gestate the resultant embryos. Therefore, the female part wants to know that her partner is a worthy male to mate with. Forsyth suggests that the presentation of a substantial packet of calcium carbonate, in the form of a love dart, may act to prove the suitor's fitness as a father—"an honest advertisement," as he puts it. In any case, the elaborate spearpoint, half an inch or so long, does act to stimulate the female gonads of the love-darted snail.

Seems drastic. Anyway, it had no such effect on me, though I was surprised. And it makes me wonder: the Romans, who so admired this animal, numbered among them some keen observers of natural history. Also, Cupid (or Eros) belonged to the Roman pantheon. Could it be that Cupid's arrow has its origin in the amatory behavior of the Roman snail? The analogy is irresistible.

The mating of snails and slugs deserves serious attention. Given another of its kind to envelop in copious slime, instead of a dry and barren finger, a snail in heat is a single-minded and devoted suitor. The wet blending of bodies that ensues strikes me as the most complete merging I know in nature. I have written of butterfly courtship as being a beautiful thing to watch, and it is. But the genteel connection they achieve seems almost chaste by comparison, and certainly lacks the passion of full-blown slug love. (Reputable zoologists write of the smacking kisses that accompany snail mating, and of the real likelihood that they undergo a neuromuscular crisis akin to orgasm.) And, lasting for minutes or hours, these molluskan mergers make the momentary matings of cross spiders (and deer) seem stinting indeed.

I was privileged to watch Roman snails couple when I returned to the same spot in Surrey this past summer. This time I kept my hands to myself and just watched. I was happy to see the great univalves present in large numbers, for it was not long before that my colleague Susan Wells and I had to place the species in the international *Invertebrate Red Data Book* due to its rarity in most of Europe. Despite its fecundity, the snail's good taste has brought its numbers to the brink of endangerment. Happily, the result of such couplings is a lot more snails. With adequate protection, the edible snail should survive to grace many more tables and nature walks.

We have no apple snails in Willapa, but we do have banana slugs by the millions. Commonly we watch them rapt in the same sort of embrace as I have described, mantle to mantle, so perfectly matched that one seems the reflection of the other, until the writhing begins. These great, shell-less snails occur abundantly throughout much of the Pacific Northwest. But if they sold as well as books of slug jokes and recipes for their preparation, they might become rare too. And if they just had a shell, they might give escargot a serious slither to the skillet.

A good-sized banana slug would furnish two to three times the flesh of the garden snail, or more like the quantity of a Roman. But most people (even those who consume oysters and escargot with gusto) would no more eat a slug than a bulbous-bodied spider. So the slug recipe books to which I referred, and they *are* a hot item, are just gags: "Slugs stick to your ribs . . . hands, clothes, shoes, etc." "Stuffed Slugs—a simple dish for simple people." And not very nice jokes, since they make fun out of excessive cruelty to what are likely fairly sentient creature (there is evidence that the exchange of love darts causes palpable pain for the partners involved). But people laugh, not only because of the novelty and the grossness of the venture, but also because of their mass misapprehension of slugs.

Slugs, in a word, are generally despised. This is unfortunate, because the native species have at least as much to offer as snails in terms of charm and interest. Also, they play a vastly important role ecologically as recyclers of decaying vegetation. Indiscriminate persecution of slugs amounts to the pointless killing of beneficial organisms. (Spiders have heard it all before.)

Slugs were almost as scarce as snails in my midcontinent upbringing. But I did know what they were. Working in my grandmother's East Denver garden, I would unearth little alien milky slugs clinging to the granite chips that formed the edges of her herbaceous borders. Picturing all slugs thus, I nearly fell off the nurse log on which I spotted my first banana slug in the rain forest of Mount Rainier National Park in 1964.

Banana slugs, commonly three to four inches long, can reach ten or even twelve inches and an inch or two in diameter—real bananalike proportions. They tend toward a yellow green with or without black spots (some populations being all black), so the name suits doubly. These magnificent animals possess sleek central ridges, pumping pneumophores that breathe air into the balloonlike lungs shared by slugs and snails, and two pairs of tentacles. The larger ones are endowed with surprisingly sophisticated eyes and olfactory organs, for perception at a distance; the shorter pair smell and feel at close range. The behavior of banana slugs involves complex responses to moisture and its absence and the ability to locate food and shelter over substantial distances. Habituation occurs: you can watch the same slug daily in the same places, season after season, year after year.

All this I watch in the wilder precincts of Swede Park, especially the alder woods where *Ariolimax columbiana* grazes lichens from the bark. Hibernation takes place in my outbuildings, where opalescent fecal swirls and slime tracks of past ventures decorate the walls and windowpanes, reminders of the slugs' gentle presence.

Other native slugs occur in Willapa, notably *Prophysaon andersoni*. This reticulated gray slug with a sulphur-yellow margin of mantle and foot can honestly be described as pretty. In certain places we find almost wholly albino populations of this much smaller slug, the size of a pencil three-quarters used. Like certain lizards, it can dehisce its tail (which appears swollen in many individuals) under attack, hence, "taildropping slug." *P. andersoni* shows a liking for wild mushrooms, especially chanterelles.

Anderson's and banana slugs show no sign of the shell their ancestors bore. But another native, less common or better concealed, suitably known as *Hemphillia camelus*, retains a rudimentary external shell into which it can no more retreat than a pony into its saddle. Though apparently useless, this atavistic hump betrays the alliance of slugs and snails in the evolutionary past.

Indeed, slugs are snails. Yet while snails are venerated or at least generally liked as being "cute," beautiful, or delectable, slugs on the whole are dismissed as disgusting and hateful. There seem to be two main reasons for this double-think. As one popular article had it, slugs "combine voracious appetites with an unprepossessing appearance, leaving behind a ravaged garden and a trail of slime."

Well. It is true that slugs do produce mucus more generously than snails, since it serves crucial roles in locomotion, sex, defense, protection from cold, and water conservation, in the absence of a shell and operculum. But what's so disgusting about mucus? Is it healthy to be so repelled by a vital substance of which we produce rather a lot ourselves? Perhaps this is just a symbol of the pathological repellency many people feel about parts of their own bodies. Handling slugs can be salutary in this sense.

One of the biological and mechanical wonders of slug slime is that is has to be slippery one moment, sticky the next, in order to effect the amazing sort of movement these creatures accomplish. True, it doesn't feel very good to the touch when it goes sticky on you. But it rubs off, and you don't have to handle slugs anyway if you don't want to. Otherwise, the appearance of slugs can be quite engaging (for they have the same appealing "faces" as snails, especially when their tentacles are out), and their colors and shapes often seem attractive to the unjaundiced eye.

As for the other charge, it is quite true that certain slugs can devastate gardens. Our own garden testifies to the fact, nightly in season, however, the damage tends to be done in the main by two or three species of unintentionally introduced European slugs: the big, black-and-gray spotted *Limax maximus*; the aforementioned little milky slug, *Deroceras reticulatus*; and most notably, the European black or red slug, *Arion ater.* The last named is the grazer that gobbles gardens in a single gulp. Large (two to five inches), deeply ridged and furrowed, and colored either rust, turd-brown, or black, the suitably named *A. ater*'s gluttony is matched only by its fecundity. Unlike the banana slug, this species dies off each winter, but not before leaving plenty of little pearls of eggs in protected places to carry on the race for the tomatoes.

But there I go, moralizing like the rest of the slug press: there really is no gluttony about it. This is simply an animal, outside its native range through no fault or design of its own, removed from its predators and parasites, faced with an amenable habitat and abundant larder, doing what it is designed to do: you'd do the same. It is not the slugs' fault, and it does no good to anthropomorphize them into willful villains. Still, we kill 'em, as I suppose most gardeners will—preferably with as little cruelty as possible, by freezing. It's merely protecting our produce.

The sad and unnecessary thing is that many gardeners and others also kill banana slugs, in fact *all* slugs—and do so with gusto, as if out of revenge for the lettuce. And there is just no point to it. For, as I've said, the natives, on the whole, stay out of the garden. When they stray in, they can be removed. It just takes a modicum of discrimination and care.

Swede Park provides a perfect example plot for slug studies. What I have found over seven years here off and on is that *Arion ater* proliferates in the garden and lawn; *Ariolimax columbiana*, in the woods; and that seldom the twain shall meet. They blend only over a narrow band of rough ground at the edges. True, an especially succulent bunch of young plants sitting in a cardboard box on the back porch one rainy night did attract every species of slug on the premises, including the natives. But the instances of banana slugs or Anderson's slugs in the garden have been relatively few. We have no battle with them. Nor, any more than for cabbage butterfly larvae, do we use any chemicals that might harm our cat, nontarget invertebrates, or ourselves: we occasionally plant dishes of beer around the periphery of the beds (death on slugs, as are the frequent beer cans in the road verges); but usually we simply pluck the caterpillars and freeze the slugs for composting, taking care to transplant the few bananas back to the banana belt.

Professor Ingrith Deyrup Olsen, a University of Washington zoologist who studies slugs, is an ardent champion of these valuable creatures. She finds them to be elegant research subjects and important elements of natural habitats, where they help to break down and recycle vast quantities of vegetation. Dr. Olsen worries about the impact of alien slugs upon the natives. Happily, at least in Willapa, I find that *Arion* rules the roads and yards, *Ariolimax* continues to prevail in the forests and woodlots, be they ever so close together. The obvious selection of disturbed lands by the black slugs and more natural by the banana closely parallels the division of resources observed by European cabbage butterflies and native veined whites. The ethnic Americans and Europeans stick to themselves, be they butterflies or slugs, along very nearly the same neighborhood bounds. Cheek by jowl, each species thrives in its chosen lands, and for the most part, only the exotics go for the gardens. But humans tend not to discriminate in their vengeance.

Too often, the popular literature doesn't help. The slug novelty books encourage a malicious and deliberately cruel attitude. And the Washington State University Cooperative Extension Insect Awareness series publication on slug control fails to differentiate between natives and others. Photographs of *A. ater* and *L. maximus* are both labeled simply "slug," and nothing is said about the existence of benign or beneficial slugs. Bananas surely may be called benign, even beneficial. The public should be educated to recognize these and other natives and to leave them alone.

More encouraging notes, however, begin to be heard in other quarters. A 1973 article in *Pacific Search* magazine, entitled "The Unendangered Slug—Ugh!" gave no strokes to native slugs, lumping them with the aliens in the nonsensical term "the slug," a meaningless abstraction, as Stephen Jay Gould calls such "the" words. "The slug" we learn, "is one of nature's most disliked creatures." But ten years later, that magazine's successor, *Pacific Northwest*, printed an article, with a different emphasis, entitled "The Lives of a Slug," by Rick Gauger. Sumptuously if not diagnostically illustrated by Dugald Stermer, the piece highlighted Dr. Olsen's work and the plight of misperceived native slugs. Banana slugs were empathetically depicted as hounded by foreign competition and domestic harassment alike. Directly confronting the earlier article, Gauger concluded that the time might come, after all, when the banana slug will be considered endangered. I don't really believe that (unless slugs take off as less-than-fast food); but I am glad to see the popular literature on slugs take a turn toward just concern and respect.

The popular impression of slugs as detestable and disgusting makes no more sense, after all, than my former hatred of spiders. For where, after all, lies the line between obsession and contempt? In which black pit of our brains forms the nauseous root of revulsion? And whence pounces phobia, to crush the impulse of interest and goodwill and inform the consciousness instead with senseless horror and needless hate?

We know that love and hate are not so very far apart, and the ways we try to show them both can be just about the same. (Why else the love darts of the snails?) When, as a boy, I reviled spiders and their webs yet exalted snails and their shells, perhaps my childish passions sprang from the same source—and ran to equally unwholesome depths. And in the foggy bottoms of my forming mind, perhaps they nearly met and might have switched around so that spiders drew my utmost yearning and snails my sickest dread.

Maybe neither was an appropriate way to regard nature. For the boundless enthusiasm of enchanted youth just cannot be sustained, while fear and loathing never please Pan. For me, coming here, obsession and aversion collided in a land where the objects of each abound; and now I feel neither. I know that at this very moment I can take flashlight in hand and inspect an enormous apricot spider repairing her great orb web or taking her rest up inside her house of woven maple leaves; or follow the silky signature of a snail's slime trail across the old plum glade to where the ivory foot and tentacles of a green snail may be seen in midslither over ripe plums on the damp ground, shiny shell in tow. Both visions would give me exactly the same reward—a solid sense of pleasure from beauty, fascination with form, and wonder at the elegance of evolution—as infatuation and phobia blend into equanimity toward all nature.

# I, Clodius

Basic questions in biogeography often tell as much about the biologist as about the animals or plants involved. For example, a record of a rare organism far out of its hitherto known range may merely reflect the travels of an intrepid naturalist, since distribution maps often record the whereabouts of specialists as much as those of the creatures they study. We call this "collector bias," or "artifact of undercollecting."

Of course the distribution of enthusiasts may prove almost as interesting as the patterns of occurrence of other species. Why does one person wind up in the thick of things, another on the outskirts?

In the previous chapter I explained my childhood obsession with snails, how in Colorado it remained unrequited, and how I ultimately made the adaptive shift into butterfly collecting. That made much more sense in light of my location: Colorado is one of the best butterfly states, one of the poorer for terrestrial pulmonates.

The irony in the tale came about thus: not many years after I changed a conchologist's hat for a lepidopterist's, I moved to western Washington State: a great place for snails, one of the worst regions in the world for butterflies. Still later I settled in Wahkiakum County, perhaps the part of the state with the lowest butterfly diversity and abundance—hence, one of the poorest patches of land I could possibly find for a butterfly man.

Now, as an author of several butterfly books, I am often asked variations on the question "What's a nice lepidopterist like you doing in a butterfly dump like this?"

The easy answer is that the lack of butterflies means fewer distractions to writing about them and other things. But beyond that, it really isn't all *that* bad for butterflies, and its relative impoverishment has its own rewards, as I shall proceed to show. However, my situation here doesn't make a lot of sense on the surface, and I would like to take this opportunity once and for all to explain how it came about and what I do about it.

It does seem logical that naturalists should choose places to live where their enthusiasms find fit fare. For example, there is the tale of Edwin Way Teale, the late Pulitzer Prize-winning and highly beloved nature writer. Several of his earlier books dealt with insects—*The Golden Throng*, *Grassroot Jungles*, and *Near Horizons* among them.

In *Near Horizons*, Teale described the insect garden in which he made many of his entomological observations and photographs. For about ten dollars per annum, he leased the insect rights to an old orchard on Long Island, and there planted "the vegetation that insects like, the things which provide the nectar and pollen and juicy leaves which delight them most." Edwin Teale's garden was an "Insect Eden." "This close-to-home stretch of grass and weeds, set among mouldering trees, has been my veldt, my tundra, my Amazon jungle." Although his job required that he live near New York City, he made sure to make his home where insects abounded.

In later years, Long Island became too developed to support the level of natural contact the writer required. So when the Teales were at last able to subsist on his freelance income, they cast about in a systematic manner to find a new home: within one day's drive of the city, yet wild enough for their countrified tastes. The successful search he described in *A Naturalist Buys an Old Farm*. One of the factors Teale took into account in selecting his new situation was that it might be a place where he could begin his insect gardening anew. A fallow farm in the rich

Connecticut woods just filled the bill. And when I visited the Teales at Trail Wood, it was obvious to me that theirs was the abode of naturalists, and carefully selected as such.

Similarly, friends of mine have situated in Arizona because they craved hawkmoths, Papua New Guinea for propinquity to giant birdwing butterflies, or Costa Rica to satisfy their passion for passionflower butterflies and quetzals. Likewise, anyone familiar with me might assume that I used to live in the Rockies because of my love of arctic-alpine butterflies. In fact, as a dependent minor, I had no choice in the matter. I was a Colorado native, and I fastened on to mountain ringlets because they were there.

Yet, as an adult, my decision to live in southwest Washington was volitional. No one compelled me to come to Willapa, nor was work a factor, since there isn't any. So why did a butterfly lover hang up his net in a butterfly desert where it rains all the time?

In earlier chapters I have described my reasons for coming to Washington—the family tales, the gravity of green, the tug of the mosses and ferns. I came as an undergraduate and (with intermittent absences) decided to return and remain. Still, I could have selected Chelan County, where my grandmother pioneer-taught in 1916, the Okanogan Highlands, or the Yakima country; these slices of the eastern Cascades abound in butterflies, rivaling Colorado's showing. When I discovered the paucity of Rhopalocera in the evergreen empire, and the concomitant richness on the drier, sunnier east side, mightn't I have settled there?

No. For one thing, it seemed that if one were to live in a hot-cold mountainous place a long way from the sea and the fernwood, one might as well live in Colorado where there are even more butterflies. The Pacific Northwest to me meant the damp, the verdure, and the maritime mildness of the coast. I still love Colorado dearly, but it's too far from the sea. So is Spokane.

For another thing, being rich, the east side has already been worked for its butterflies. John Hopfinger in the North Cascades and the Okanogan,

E. J. Newcomer in Yakima and the South Cascades—these longlasting pioneers prospected the butterfly lodes of these territories for the first half to three-quarters of this century, making many spectacular finds and filling in the form books for eastern Cascades butterflies. Of course, they didn't find them all, and energetic collectors of today and tomorrow will continue to improve the picture dramatically. I much enjoy visiting the east side during spring wildflower time, and we conduct an annual butterfly count in the Chumstick Mountains near Wenatchee. Without doubt, there's some fine butterfly country over there.

But something in me desired *terra incognita* for a homeland, vis-à-vis butterflies. And if Willapa was anything, it was that. As I speculated in *Watching Washington Butterflies* in 1974. "Still, I can't help but think there must be biological pockets yet unsampled somewhere back in the Willapas, on mountain balds, flood plains, and so on. In searching for them we will find a few special things, if we are lucky or skilled."

So, ironically, it was partly the search for butterflies that drew me to this relatively butterfly-free location. When I was a graduate student at Yale, my thesis research involved an analysis of the butterfly distributional patterns of Washington relative to nature reserve occurrence. Acting on that 1974 speculation, I visited the Willapa Hills during the summer of 1975 to fill in a gaping lacuna in my data. It rained, and I found none: the gap remained and seemed to broaden by comparison with our knowledge of the rest of the state. But at least I saw the valley that was to be my future home. My friend Denny Gillespie, then a historical preservation and recreation specialist for the Cowlitz-Wahkiakum Governmental Conference, took me to see the covered bridge at Gray's River and other historic structures in the area, in lieu of the damp habitats we'd hoped to explore. I was quite taken with the countryside. Even in its sodden state, I thought it would be a nice place to live.

Subsequently, in 1978, I found myself working for The Nature Conservancy in Portland, Oregon. On weekends, I frequently crossed over into Washington to carry on my studies of that state's butterfly fauna. During my doctoral survey, Wahkiakum County was the only county in the state for which I found no butterfly occurrence records whatever (hence the 1975 attempt). Later I learned that E. J. Newcomer had made one visit in 1962, found half a dozen species flying up the Elochoman River valley, and turned around to head back to the greener pastures of his Yakima canyons, where one hundred species might be found in a season. Still, my colleagues and I—the Evergreen Aurelians—knew next to nothing about Willapa butterflies.

The six of us who banded loosely together to map and study Northwest butterflies took our name from "The Evergreen State" and "The Aurelians," the first entomological society, which flourished in London in the eighteenth century. We set as one of our goals the gathering of as many distribution records as we could from all over Washington, Oregon, Idaho, and British Columbia. So for my part, I kept plugging away at southwest Washington, looking for weekend windows of opportunity to head north across the river between the rains and never doubting that there would, after all, be some butterflies to find.

So it was that one Saturday in August 1978 my field assistant, David Shaw, and I packed sandwiches and nets and set out for Wahkiakum County, one hundred miles distant from Portland. The sun stuck with us. We took the last ferry on the lower Columbia from Westport, Oregon, to the river's largest island. On Puget Island, a green lozenge of farmland plunk in midstream, we found red admirals, painted ladies, Milbert's tortoiseshell, and orange sulphurs, all new for the county if common elsewhere. Encouraged by these modest discoveries, we continued over the bridge into Cathlamet, county seat of Wahkiakum, headed up the Elochoman River road and found the same species that Newcomer had tallied there sixteen years earlier: purplish coppers, mylitta crescents, veined whites, Clodius parnassians, and western and pale tiger swallowtails. Not much for the Cascades, yet this bag was more than we expected in the logged-over, rainy, moldering Willapas.

Heartened by the unexpected bounty, we decided to head westerly, with no adequate maps. There was no more collecting as the afternoon hours were spent negotiating a crossing of the Gray's River Divide, on logging roads, that in later years I would learn was rather tricky. We made it, rolled down the Gray's River Gorge and across the bridge described in "Robert Gray's River," and struck State Route 4 just east of Gray's River village.

It remained to turn left, head up the Columbia in the gloaming, gloating over our historic catch, and so back to Portland on the interstate. But as I began that turn, I suddenly lurched right onto the highway instead. "What's up?" asked David Shaw. "If I remember rightly," I replied, "there's a fine old covered bridge up here a short distance. Might as well show it to you while we're here." From New York State, which has rather a lot of covered bridges, and hungry, David nonetheless yawned and said, "Okay."

Show Shaw the bridge I did. And as we returned over its plank approach, rounded a curve by a grand old barn, and rose toward the highway on a narrow country lane, we passed a striking white house on a hill—an old, plain Victorian two-story farmhouse, surrounded by huge hardwood trees. In front of it stood a sign: "For sale by owner." "That's where *I* want to live," I suddenly decided and announced. I turned the VW bus into the stony drive between immense overhanging oaks and pulled up to the house. And that, leaving out a few subsequent details, is how I came to live at Swede Park, Gray's River, Wahkiakum County, Willapa Hills, southwest Washington—*not* the butterfly capital of the West.

Of course, it took a while to sever city ties, trump up faith in the freelance life, and move out to the country as I had long wanted to do. (Unlike Edwin Teale's, my writing career did not yet justify such a move.) And it took a while after that to find out just how subtle were the butterfly charms of Wahkiakum. But I already had an inkling from my explorations so far. I knew I would be settling into no tropical butterfly paradise; I can't claim it was any surprise. I knew roughly what I was in for, and I had meandered through enough meadows elsewhere in western Washington to be ready for the worst. In that frame of mind, perhaps I exaggerated my plight; and in fact it hasn't been so bad.

Friends from the Midwest or the Southeast or Colorado or Connecticut or practically anywhere else wonder openly how I come to terms with the paucity of butterflies. (Only western Washington and Maine are thought to be so poor.) And yet, I tell them, there's plenty to do with butterflies here, even though, in the first seven years, I recorded only around thirty species in the county. (I found more than twice that number along a single old ditch east of Denver in my youth.) And there are certain attributes of a small fauna that recommend themselves to the watchful naturalist.

For one thing, as I mentioned, a modest assortment of species is a palpable, tenable array that an average intellect can grasp. For another, a simple fauna allows one to concentrate on the individual animal. In the tropics, even in Colorado, with many species and individuals of butterflies, the individual tends to get lost among the many. Here, in contrast, where a couple of dozen individuals of five or six species make up a reasonable day's tally, one has the leisure to pay full attention to singletons. An autumn walk to the mailbox, for example, may take half an hour, as the cat rolls on the warm road for a tummyrub and I squat watching a lone mylitta crescent nectaring on a hawkbit or basking on a leaf. It may be the only butterfly I see all day, but I'll see it well. Or a female purplish copper may come along, and I'll put the crescent up to investigate her. The flickering of eight orange panes, overlain with spots and lunules of black, will have enriched my morning as they rendezvous, then part.

Such an argument may not impress the collector who aspires toward a bulging bag of specimens; but the butterfly-watcher (which I chiefly am now anyway) has much to learn from it. A properly observed butterfly gives much greater pleasure and understanding than one merely ticked and tallied, or netted and pinned. Most canny collectors know this already and spend as much time peering at their prey as swinging their nets. I have learned that lesson here—where an encounter with a common creature can make my day, where there are no "junkbugs," and where no butterfly is to be taken for granted.

Consider Clodius, for example. The Clodius parnassian (*Parnassius clodius*), related to the European Apollo butterfly, was named not for the Roman emperor of the chapter title pun but for a Roman tribune in league with Caesar, or perhaps his sister. Swallowtails, which are in the same family, were frequently named after Roman gods or heroes in the Trojan war. Other parnassians received godly names—Apollo, Phoebus, Mnemosyne—but the Russian author Ménétriés, for some reason, felt like commemorating one or two people with decidedly mortal qualities: Clodius was an ephemeral Roman politician described as a gangster and an opportunist; his sister Clodia was said to be a woman of highly unconventional morals. However, both brother and sister were considered beautiful; in fact they bore the surname Pulcher. Perhaps this fact led to the ultimate immortality one or both of them achieved in the naming of this rather magnificent butterfly.

The Clodius parnassian is a pale, waxy, white-winged insect. Quite large, spreading up to three inches, the wings are dotted with ink spots off a thrush's breast and cherry drops of red. Both larger and more translucent than the males, the females lack scales over much of their wings, making them largely transparent. Additional characters include short, all-black antennae, a compact black body that is furry in the males, naked in females, and small eyes compared to the swallowtails to which they are related in structure if not appearance.

The yellow-spotted, black velvet larvae apparently mimic cyanide-producing polydesmid millipedes. These caterpillars feed on wild bleeding heart, an attractive, leafy herb with rose-pink, pendant flowers built of petals fused like the lobes of a heart. Clodius and its host make one of the great aesthetic partnerships of the Northwest woods: the elegant green stands of lacy bleeding heart, flecked with aromatic pink flowers, frequented by the great, floppy white butterflies, speckled with scarlet and jet.

Females deposit their eggs among the bleeding-heart stands, and sometimes on shrubs whose leaves will fall to the ground where next spring's host-plant foliage will emerge. After hatching in early spring, the young larvae feed secretively and molt their skins several times. Usually by May Day they will have attained full size, and shortly thereafter they will pupate. Parnassians are unusual among butterflies in that they form their chrysalis within a silken cocoon such as many moths employ. That of Clodius is strong and well formed.

David McCorkle and Paul Hammond have studied Clodius in some detail and they consider it to be a primitive butterfly in evolutionary terms. The courtship behavior seems to reinforce this idea. Lacking any refined courtship pattern such as more advanced butterflies possess, male and female Clodius get together in the bluntest, simplest manner. The male simply forces the female to the ground and copulates with her, in a type of mating referred to as rape behavior by ethologists. After coitus, the female bears a genital blocker called a sphragis that prevents further matings. This biological chastity belt serves to ensure that the first male to mate with a female will pass on his genes. Formed from a male exudate, the white, waxy, hooked shield borne by mated Clodius females is the largest and most elaborate sphragis of any parnassian species. McCorkle and Hammond believe that the generalized habitat, rape behavior, large sphragis, and well-formed cocoon all demonstrate the primitive condition of Clodius and its near relatives. More advanced species are thought to have left these characters behind in favor of more refined techniques and structures.

Clodius truly is a creature of the Northwest. It occurs in Yellowstone and Yosemite but concentrates up through Oregon and Idaho into Washington, British Columbia, and southeast Alaska. It has its "metropolis," as the old-boy lepidopterists would write, west of the Cascades, and it can be abundant in the damper parts of the Cascades, Olympics, and Willapas. Clearly, Clodius has come to terms with the moist conditions that somehow seem to exclude or reduce so many species. The closest thing to a desert population feeds on a white-flowered bleeding heart in the wetter clefts of the Snake River country in southeastern Washington and Idaho, where it was discovered by and named for one of the Evergreen Aurelians, Jon Shepard. (Jon and his wife, Sigrid, are *Parnassius* authorities.) Another Aurelian, Dave McCorkle, first found bleeding hearts to be the larval hosts of Clodius; prior to that it was thought to feed on stonecrops (like other parnassians) or violets.

*Parnassius clodius shepard*i tends to be larger and lighter than other populations. But all of the races of parnassians vary greatly and run together, leading me to call most of them by their species names and to ignore the many subspecies names that have been applied. Our only other parnassian was called Phoebus, another name for the god Apollo, meaning "brilliant," when we thought that our second parnassian was the same species as the Old World *P. phoebus*. Jon Shepard has since shown that the proper name for it is *P. smintheus*—yet another name for Apollo, from the *Iliad*, meaning (inscrutably) Mouse-God.

The Smintheus parnassian is smaller than Clodius, bears red spots on the forewings as well as the hindwings, and has white-ringed antennae. Smintheus tends to occur higher in elevation (well up into the arctic-alpine zone) and may yet be found in the highest fastness of the Willapa Hills; while Clodius occurs from sea level up to the middle elevations where Smintheus takes over. Way north in British Columbia, Alaska, and Siberia, a third North American species occurs. Named for a mere mortal of a Russian, *P. eversmanni* possesses yellow males and great mystique.

During the long days of May, Clodius is but a stuffed husk wrapped in a cobweb, tucked among the duff where last year's bleeding hearts rot. But come June, the bramble blossoms bloom and so blooms the butterfly: a red, white, and black package hanging from the golden stamen bundle, probing the blackberry blossom as deep as it can reach for the sweet nectar. White wings glide among the pallid parasols of the plants, *blanc* butterflies flapping through the garden and across every forest road, path, and clearing.

I recall one female I spied in the act, her body dusted with pollen, so heavy with eggs that she pulled the flowers nearly to the ground as she clambered among them. I feared that she might impale herself on a bramble thorn like a thick, stubby insect pin under her own weight. But she didn't, and eventually she took wing in search of bleeding-heart foliage for her eggs, as if treading water on the hot summer air, so fat in her fecundity she could barely fly.

Such are the sights to be seen in examining a single species. And I speak of the shallowest, most casual observations, aesthetically and emotionally fulfilling but intellectually simple: often the only kind we have time or talent for. Imagine, if you can, the adventures in basic biology to be had with Clodius alone were one to pursue them with invention and rigor.

Many people collect insects as children; in fact, almost everyone probably does at one time or another. Some few keep it up and become entomologists or else leave the collection behind but remain serious watchers. And a very few of these, through the desire to know more, take up the tools of basic investigation, apply the scientific method, and, through research, contribute to our overall understanding of the microfauna. But the actual number of people trained, equipped, and otherwise prepared to conduct even elementary research is vanishingly small compared to the beginning cadre of casual collectors. That is not only okay, it is inevitable. But imagine—it leaves open a field of tremendous breadth and opportunity for the few who persevere.

Several of my English entomological friends, among them Jack Dempster, Marny Hall, Jeremy Thomas, and Martin Warren, have developed tedious techniques for the study of butterfly autecology—the ecology of individual species—whereby a richly detailed picture can be assembled of all the factors involved in the life and death of organisms. Paul Ehrlich's research team at Stanford University has fashioned one of the most sophisticated of these species-portraits, working with Edith's checkerspot butterflies around the West. No one has done any of this in Willapa; so even with a limited fauna, the horizons are broad for the watchful worker or intelligent watcher.

With Clodius alone, we lack answers to a thousand questions. Among them: how, when, and where do the caterpillars feed? What makes a suitable bleeding-heart stand? How many survive, and what are the chief causes of mortality? What is the winter survivorship, and where does pupation take place? How large and mobile is a population, and with what climatic or ecological factors does it fluctuate? How far do the adults fly in their lifetimes, over what paths, and how do they find their mates and host plants? What are the chief adult nutrients? What is the natural range of variation, and how much of it is genetic versus environmental? How does the female find the host plant, and exactly how is her sphragis formed during mating? Do the caterpillars really mimic the millipedes, and how does their behavior reflect this?

Dave MeCorkle's experiments with mice, millipedes, and caterpillars suggest true mimicry going on here. Indeed, McCorkle and Hammond's work on the species has thrown light on many aspects of its biology. But an enormous amount remains to be learned about this and all butterflies.

Of course it is easier to sit back with a beer and watch a fat female suck blackberry nectar, and rhapsodize about the pleasing juxtaposition of form and color. I enjoy that. And yet I also like to contemplate these possibilities, in case my situation should ever permit or encourage a return to rigorous scientific investigation; to remind me to be watchful and to record my observations carefully (too many of us squander our observations in the dime store of our memories); and to demonstrate the range of research available with a single species in butterfly-poor Willapa.

And there are others. Our two most abundant spring butterflies, the spring azure (*Celastrina "argiolus"*) and the veined white (*Pieris "napi"*), both beg understanding. I place quotation marks around their specific epithets because both belong to Holarctic species groups whose status remains obscure. Very likely our veined whites are not the same species as the European green-veined white, nor are our spring azures likely to be conspecific with the holly blue of Great Britain. Yet, for lack of sharper intuition or evidence, we give them the same scientific names. Clarification awaits much greater insight based on solid knowledge of the basic biology and genetics of the various populations involved: not just how they differ in spots and size in the specimen trays, but also how they use their host plants, arrange their generations, behave to prevent hybridization in nature, and array themselves across the countryside, as well as how evolution has caused their genitalia, chromosomes, DNA, and protein enzymes to differ.

The full picture will require integrated research efforts for clarification, many of them quite complex and involving sophisticated laboratory equipment and personnel. But other elements, equally important, await investigation by any careful, observant person willing to invest the effort and thought. So what that our Washington whites and azures may be the only butterflies on the wing for many days and in many places in spring? So much the better for concentrating on them. It can be difficult settling in on a single species project, with dozens of other varieties impinging on your sensibilities at every turn.

Aside from basic biology, another line of inquiry suits my situation here very well indeed. This is butterfly biogeography: why do butterflies occur where they do, and (perhaps more to the point in western Washington) why do they fail to occur elsewhere?

Biogeography has to do with the accumulation of distributional data, its display and interpretation, and construction of theoretical models to explain plant and animal distribution. Like much of natural science, it involves the erection of generalities through the examination of specific cases. The basic datum of research in biogeography is the occurrence of organisms at given sites. The pursuit of such data comes under the term faunistics for animals, or floristics for plants. A purely faunistic approach is not very fashionable in zoology any longer, since it lacks the glamour and potential of analytical mathematical studies. But faunistics and floristics remain important, since before you can analyze, conserve, manage, model, or do anything else with animals and plants, you have to know their whereabouts.

Of course, the presence or absence of organisms in particular sites depends upon their personal traits and those of the sites themselves—the *ecological* relationship between living things and the land. Ecology has everything to do with the specific places in which animals and plants may be found. (This is why habitat conservation proves so very important to the maintenance of rare species.) The broad patterns of distribution, on the other hand, result from a number of biogeographical factors that are not strictly ecological in nature. These include the mobility of organisms, physical barriers to and corridors for their movement, climatic history, continental drift, mountain-building and erosion, desertification, sea level changes, and the spread or contraction of other species, including (especially) human beings. So the location of wildlife describes a pattern born of many causes, both physical and biological, current and historical.

Understanding these patterns begins with the gathering of records of occurrence in a given time and place. For my thesis study, I accumulated some ten thousand butterfly records for Washington. Analyzing these, I was able to propose and support butterfly provinces for the state, regions distinct from one another based on their substantially different butterfly faunas. While ecology plays a role in their makeup, these provinces were primarily biogeographical in nature rather than ecological. In other words, they are defined in terms of their affinity with other regions over the long term, rather than by their present-day vegetational features. Comparing the proportion of land designated for biological conservation in each butterfly province enabled me to show that certain highly distinctive parts of the state had been neglected in the land-protection process. I am pleased to report that several of these lacunae have since been filled and that the butterflies helped do it.

Since then (1976) the Evergreen Aurelians have carried on the task of aggregating records. My good friends Jon Pelham in Seattle and John Hinchliff in Portland monitor and record Washington and Oregon records, and we now have in excess of fifty thousand data. These records are being shared with Oregon and Washington Natural Heritage and Nongame programs. This will let us print out detailed maps of distribution, and in turn to develop better pictures of each species's distribution and conservation needs. The Xerces Society, an international invertebrate conservation group, hopes to use our Washington and Oregon efforts in this direction as a model to be encouraged nationwide.

I founded Xerces in 1971 while studying butterfly conservation with John Heath and other British mentors at the Monks Wood Experimental Station near Cambridge. John developed the British Butterfly Recording Scheme, part of Her Majesty's Nature Conservancy Biological Records Centre, which he directed until recently. Probably the world's most sophisticated system of applied biogeography, BRC prepares atlases of the flora and fauna of the British Isles based on records contributed by thousands of amateur naturalists. The data, checked for accuracy, are displayed with dots on ten-kilometer-square grid maps. *The Atlas of Butterflies of Great Britain and Ireland* presented the results, giving naturalists and conservationists in the United Kingdom an extremely useful tool.

When I returned home, I brought with me more than a wish to see the successful British recording scheme emulated here. One lesson I learned well from John was the pleasure to be had in "square-bashing": seeking new records for hitherto undotted grid squares. On off-days or on the way elsewhere for meetings or fieldwork, we would often visit poorly recorded squares and tick species present. Not only does this make the maps better, but each new dot represents a fresh quantum of data—a scintilla of fresh scientific information about the world we live in. It is an easy, pleasant, and satisfying way to enhance the sum of human knowledge about butterflies, or any other organisms. True enough, this is only alpha-level observation, with no pretenses of being on the "cutting edge" of biology. But you've gotta know where something is before you can ever get to beta or gamma research.

Having become addicted to butterfly square-bashing in England, I've carried on over here. I've mentioned the Evergreen Aurelians' record-gathering operation. Our "squares," rather than ten-kilometer grid squares, consist of U.S. Geological Survey map quadrangles, Township/Range survey sections, or tenth-of-a-degree latitude-longitude rectangles. Any of these makes an eminently bashable unit, which can be further subdivided to suit the desired degree of resolution, or to increase the challenge and rewards for recorders.

Colleagues of ours elsewhere use other systems. Ray Stanford keeps the western states' records on a county-by-county system. Paul Opler likewise employs counties as his recording unit for the eastern states. Kenelm Philip directs the Alaska Lepidoptera Survey; he bashes all of Alaska, the Yukon, and eastern Siberia, when he can get there, on a point-by-point basis. That gives him an infinitude of possible dots! My local self-assignment is rather more limited.

Even so, by doing nothing more than identifying the butterflies I see and recording the data with exactitude, I can make a sound contribution to Washington biogeography. And this is one of the very real reasons I have been more or less content with the paucity of butterflies in southwest Washington: you don't see many, but when you do, they tend to be interesting, and very often they represent new records at the county or section level. And that is satisfying. That everpresent possibility of making alpha-level discoveries is one of the real attractions (for me) of living in an area poorly worked by previous naturalists. I can go out any time and enjoy the prospect of finding creatures never before recorded from these hills, and not infrequently that's just what happens.

When I came here, just half a dozen species of butterflies were known from Wahkiakum County; now the list runs to five times that many. It could, with effort and luck, be pushed to fifty—so the challenge will remain for some time to come, and actually it extends to this entire corner of the state.

A county record is a buzz, but local discoveries can be more than that. Some real surprises occur. A decade ago, casual collecting turned up the Oregon silverspot butterfly near the coast in Pacific County, on the western edge of Willapa. This find represented the rediscovery of a federally listed Threatened Species, previously thought to be extinct in Washington. Another time I happened to net an anglewing beside the woodpile in my own backyard, and it turned out to be *Polygonia oreas*. The oreas anglewing, russet above and nearly black below, is the rarest and most mysterious comma butterfly in the state. This was only the second catch for the Willapa Hills; I'd made the first up the Elochoman River a summer before. This year I added another anglewing, the faun, as well as the silvery blue and the hydaspe fritillary to the county list, and next year I will especially seek the spring forest hairstreaks and dusky-wing skippers. As I add new red dots to my maps and charts, I take real pleasure in the simple discoveries they represent.

Yet county records mean little in and of themselves, beyond the gut thrill of the find and the extended understanding of range. The greater value and challenge lies in the interpretation of the patterns they convey. Biogeography can be a frustratingly imprecise science, since one can reconstruct pasts only hypothetically. The whereabouts of rivers and glaciers and sea levels, the impact of earlier eruptions of Mount St. Helens and other Cascade volcanoes, the extent of warming and drying trends—these things can be inferred from geological evidences and pollen records, and they help. But in the end, you have to make some informed guesses.

The study of butterfly distribution in the Willapas gives me a chance to ask some basic questions. I've often joked that my life's work is to explain the depauperate nature of the western Washington butterfly fauna (and it does need explaining). In 1974, I wrote in *Watching Washington Butterflies* of the Willapa Hills, that this "oft-posed question . . .

might as well be addressed here as anywhere." I still think that's true. By looking at the limiting factors for local species, we can begin to view the larger issue.

Take our only two local satyrs, for example. A refinement of the irony of my situation lies in the fact that my favorite butterflies are the browns or satyrs, subfamily Satyrinae. Well represented in both the Colorado high country and the English lowlands, they are nearly absent from the maritime Northwest. One alpine (*Erebia vidleri*) and one arctic (*Oeneis chryxus valerata*) dwell in the heights of the Olympic Mountains, to the north. One other arctic follows a strange disjunct pattern: *Oeneis nevadensis* occurs in the Cascades, on Vancouver Island, and the San Juan Islands in Puget Sound, skips the Olympics and Willapas, then shows up again on Saddle Mountain in the Oregon Coast Range a few miles south of here. It is probably left over from an earlier distribution when it was widespread across a drier region. Such an organism is called a relict and is said to display a *relictual* pattern. Anyway, it isn't here.

Only two satyrines enter the southwest sector of the Evergreen State, the large wood nymph (*Cercyonis pegala*) and the northwest ochre ringlet (*Coenonympha tullia*). Both species follow the Puget (glacial) Trough down to and beyond the Columbia into the Willamette Valley. They extend westerly in the Chehalis and Willapa valleys respectively almost to salt water, and both species fly near the coast in Clatsop County, Oregon, across the Columbia. In other words, Wahkiakum County is surrounded by satyrs on three sides, yet seemingly nurtures none within its boundaries—except that wood nymphs duck into a few drier spots around the edges.

In fact, a number of species drop out just north or east of Wahkiakum, or in its eastern reaches. Sara's orange-tip, for example, occurs abundantly over most of Washington, except close to the ocean. It flits brightly along the river shore east of Cathlamet, but scarcely a foot farther west than sunny Nassa Point. Other organisms—butterflies, trees, shrubs, flowers—observe the same blind border. Why? Why can't I find ochre ringlets at Swede Park, when the habitat seems just right and they occur virtually all across the west, including adjacent counties on all sides?

Both the ringlet and the nymph feed, as larvae, on grasses of various species. They accept alien as well as native species and occur in disturbed as well as natural habitats. Seemingly suitable territory for each abounds throughout the county, yet the butterflies remain no-shows. And so it might be said for western Washington as a whole: many is the collector from elsewhere who has viewed our expansive, flowered meadows with high expectations, only to turn up little more than a handful of cabbage whites—if that!

We mouth excuses about heavy forest cover, the youth of our flowery open spaces, gray days, being rainier than thou—and reckon it must be some combination of the above. The clear-felling of the forests will have caused some woodland species to be rare. But that doesn't account for the iron curtain that satyrs seem to face around my neck of the woods, when they proliferate in the Puget lowlands.

Nothing obvious to the eye explains the pattern. But when one plots rainfall figures against satyr distribution, a possible answer emerges. All of the stations for wood nymphs and ringlets receive average annual precipitation of eighty inches or less, whereas Gray's River gets 116 inches on average. It turns out that a number of species occur roughly east of a line from Vancouver through Seattle to Portland described by Interstate 5; west of there, they make excursions up drier valleys into favorable locations (such as the south-facing, sunny banks of the Columbia east of Cathlamet, where Sara's orange-tip flies) into rain shadows such as the San Juan Islands, and along agricultural corridors that cross the wet forest barrier, such as the Chehalis. Very few species are truly coastal in distribution, and not many more occupy the wetter central massif of the Olympic Mountains or the Willapa Hills.

So it would seem to be rainfall that excludes the satyrs from Wahkiakum, while allowing them to penetrate more westerly yet drier realms. Much of satyr-studded Britain (eleven widely distributed species) resembles southwestern Washington in aspect, yet its rainfall seldom exceeds forty or fifty inches in most parts. The only butterflies that seem to do well here in the rain forest are those truly adapted for very moist conditions, like the spring azure, veined white, Clodius parnassian, and woodland skipper, while a few others get by.

And yet, rainfall can't be all, because the true rain forests of the tropics—often much wetter overall than here—host vast numbers of species (over one thousand kinds of butterflies have been recorded from a small preserve in Peru, for example). But the tropical rain forests are warm as well as wet. While Willapa is mild, it is cold by comparison. My rising hypothesis suggests that a combination of damp and cold conspire to drive butterflies out of Willapa: or at least to exacerbate the effects of the heavy forest barrier, the relative lack of sunshine due to cloud and shade, and the biogeographic isolation that likely depressed the numbers in the first place. And this forms my working model for the butterfly poverty of Willapa and of western Washington.

One additional effect may be seen (and regretted). Species on the edge of their range tend to be more specialized (hence restricted) than those in the heart of their homeland. Many of the butterflies that manage to occur here do so by liege of the seasons, barely making out, so they have few colonies, tend to die out locally and rely on frequent recolonization, and occur normally in modest numbers. So a species quite common elsewhere, such as the mourning cloak or Milbert's tortoiseshell, may show up only once or twice a year around here. It can be frankly frustrating to have a species on hand, its required resource in abundance (willows and nettles in the case of the two mentioned), yet for them to persist in being so uncommon as to be next to absent.

I have come to rely on what I call the Rot Factor as my chief scapegoat. I believe that the greatest difficulty in survival comes in winter, when egg/larva/pupa (or, with the cloak and the tortoise, adult) must last through the long, wet months, alive. Litter-hibernators risk flooding for weeks on end; leaf-resters are subject to soaking just the same by near-constant rain. In the warm tropics, diapause isn't necessary and generations turn over quickly, back-to-back, never having a chance to rot in the rain. Here, I suspect, winter rot may account for most of the mortality and much of the rarity and outright exile of butterflies. My friend Jon Pelham has coined a term for those few species that have adapted to withstand the Rot Factor: mold butterflies.

The other side of the scale is that a good, bright, warm, dry year, or a succession of them, can reverse the situation rather rapidly, sending species well beyond their former ranges, swelling numbers of residents and immigrants and providing constant surprises. Such a year was this (1985), following on the heels of two pretty mild ones before it. ("I doubt if the next century will see another three-summer decade like this," a lifelong resident Finnish logger told me recently.) Butterflies responded beautifully to the relatively dry winter and spring and the long, sunny summer. While I was in England, kicking insects out of the sodden turf for a butterfly-watching tour, Jean the Postmaster reported greater numbers of tiger swallowtails here than she'd ever noticed before—a hundred or more in her garden at once, she reckoned.

Upon our return, we found Clodius on the wing in fine numbers. On my birthday, we conducted a butterfly count up Hull Creek during which the first fritillaries ever recorded for Wahkiakum County appeared—abundantly! And the hydaspe fritillaries continued to show into September, even appearing in the garden. Sleek, black-and-white banded Lorquin's admirals showed in quantity for the first year ever, as did anise swallowtails, weighing down the sweet williams and phlox with their big wings. Well into October the woodland skippers continued to throng the asters and the butterfly bush by the score, eight at a time on one small sedum, zipping out for a sip and a bask every time the sun sneaked through the autumn showers, until the first hard frosts finally ended their days.

One of the biggest surprises came when five-year-old Katriina Ervest, junior member of a 4-H entomology club, daughter of the logger mentioned above, snared a large wood nymph near the mouth of Deep River—the first one in western Wahkiakum County. That's just the sort of expansion one expects to see in a favorable year, although it's a big one for so sedentary a species. I suspect it will die out with the rains. Or perhaps it will take; or maybe it's been there all along, in numbers so low we never noticed it. If so, I'll have to take my rainy-day satyrs theory back to the drawing board. In any case, it gives hope that we may yet have satyrs at Swede Park, especially if the trend of sunny summers goes on. One thing being a butterfly lover out here tends to do is make you an optimist: if hope didn't spring, you probably would—right out the window.

In all probability, however, the rains will resume and exert their soft rigors upon the land and its creatures. Yet from time to time, we will have this sort of a year to experience. Or one such as 1983, when (following next to no winter at all and precocious sunny days) spring azures flew in February, and the veined whites came on in such numbers in April as to whitewash the salmonberry brakes.

As with the very sun itself in this part of the world, so with butterflies: there is so little to go around that when it appears, we truly appreciate it. Such full appreciation is less easy to feel in southern California.

With the proper attitude, butterflies need not be boring here after all. How could they be boring, when a prospective peek into an alder glade reveals a glossy dun skipper, far outside its previously known range? When colorful Clodius flops its ghosty form over sweet-scented patches of bleeding heart? When three species of swallowtails may all at once attend the nectar dripping from the mauve rhododendrons at the edge of the lawn? Or when a red admiral, favorite of my boyhood hunts, crimson-banded black insect of improbable beauty, probes the juices of fallen apples fermenting in the October sun; then flies up onto the whitewashed house to take the rays one more time before the frost? These and many other images spring to mind in my less-than-a-decade at Swede Park—each an *individual* image, free from competition. As our butterfly garden matures, there will be more.

And that's just butterflies: there are also the moths, ten or twenty times as many species and scarcely studied in the entire Coast Range. And the rest of the insects, some of which (like flies, beetles, and wasps) are richly represented here in terms of both species and individuals. But even a word about these will have to await another chapter in another book, after I've had a chance to broaden the bounds of my entomological knowledge, like Teale, in my insect garden.

I might just have that chance, for the butterflies here will give it to me. I remember Oomsis, New Guinea, with its blue Ulysses swallowtails and green and golden birdwings; Atenas, Costa Rica, teeming with neon preponas and rainbow longwings, one hundred species in a day; and Gothic, Colorado, realized butterfly dreamworld of my adolescence. Sure, I sometimes long for their riches. On dull, damp days, or sunny ones with nothing on the wing, I've been heard to mutter, "I'm gonna move somewhere with some goddam *bugs*!" But my threats ring hollow. It's not such a bad spot for butterflies after all. And in its very modesty as a butterfly place, it is a good place indeed to become a better, more rounded naturalist.

At the tail end of the last butterfly season, an American painted lady came to the little butterfly bush in our butterfly garden. My stepkids, Dorothea and Tom, spotted and netted it. It was a new record for the county and the Willapa Hills, and the first I'd ever seen in Washington. Rather worn, it bore the blemishes of a long flight from who knows where. But a week later, a second *Vanessa virginiensis* appeared on the *Sedum spectabile*. Perfectly fresh, the pink on its ventral forewings matched in intensity that of the plant on which it nectared, and its big eyespots stared with blind but brilliant blue pupils. Obviously it had emerged nearby and recently, leading one to hope that there might be more. Big deal, some might say, who come from where it is a common sight. Big deal indeed.

# Waterproof Wildlife

If the rain repels humans and butterflies, it well suits the newts and certain other animals. I call these the waterproof wildlife of Willapa. For those less well prepared, the point is adaptability rather than true suitability. If a duck takes to water like a duck, then a sapsucker may do so somewhat more reluctantly but do so nonetheless, rather than abandon a fine, sappy old orchard in a downpour. The key word is tolerance, and cheerfulness has very little to do with it. Whether naturally suited to water or not, an organism must in some way come to terms with the fact of much wetness if it is to survive in Willapa.

Actually, it's easy enough for us. As David Brower said in the narration to a Sierra Club film on the North Cascades, we do well to remember that skin, after all, is waterproof. With a bit of Gore-Tex or oilskin, and a few adjustments to prevent hypothermia and severe depression, humans adapt to more than one hundred inches of rainfall annually quite readily.

It can be more difficult for other animals. You might think, for instance, that slugs in their shiny damp viscidity were better at being wet than we are. But slug skin is not waterproof. These walking water balloons can drown or desiccate quite easily if conditions exceed their tolerance for moisture or drought. So, while slugs prosper in the rain world, they do so through sensitive physiological and behavioral adaptations to moisture excess. They are able to find the damp and stand it with equal proficiency.

Quite a few creatures have managed the task and some have mastered it to the point where we really can call them waterproof wildlife. Among the best are the newts and their relatives, the Amphibia. Like the fish from which they evolved, amphibians retain their physiological dependency upon water. In the larval stage most possess gills, either internal or external, and therefore are necessarily aquatic. Adults may be either with or without lungs, in the latter case breathing through their skin. That trick requires thin, moist skin, able to lose or gain water readily, and, consequently, damp conditions. "Drinking" too is by absorption through the skin. Clearly, these creatures find the water fine.

So amenable do amphibians find life in the wet world of Willapa that we can quite properly call it salamander land. According to Dr. Dennis Paulson, zoologist at Tacoma's Slater Museum, Wahkiakum County is probably the center of diversity for Washington salamanders. This means that more species occur here than anywhere else in a state known for its salamander diversity. A glance at the field guides shows more species in the American Southeast; but the Southeast is essentially subtropical, and many groups have proliferated there in the ancient, moist hardwood forests and coastal plains. For a young, far-from-tropical region, our salamander fauna is impressive.

In little more than casual searching, my family and I have found red-backed, Olympic torrent, Dunn's, northwestern, and long-toed salamanders in local woods and streams, as well as the ubiquitous rough-skinned newts. Ensatina, Pacific Giant, Van Dyke's, and the others will probably show up in good time. We search in the classic places—under logs and stones, alongside clear streams, on mossy forest floors, often with good results. Salamanders make good terrarium subjects, though we usually release them before long in our own stream, whose ravine contains all the above-mentioned microhabitats. So far, two or three species have "taken" or else shown up on their own.

With their aquatic breeding habits and larval stages, and the need to keep their skins moist, salamanders perfectly suit the land of much rainfall—or vice versa. Suitable habitats, as we have found, abound. Sometimes anthropogenic settings suffice for, or even attract, the amphibians. A swimming pool valve-box on Orcas Island proved a treasure chest of trapped newts and salamanders, which we released to a nearby pond. Recently a young northwestern salamander appeared in our stone cellar, having arrived through the walls like the winter spurtings of the floodwater, or up the sump-pump hole, but now in late summer nearly dried to death. It too was saved to enrich our pond, once for trout and now a sanctuary for salamanders.

That find put me in mind of last March, when the spring rains brought many amphibians in search of one another out of hiding and onto the slick roads. One can see the attraction—the flat surface of the highway would be suitably moist, yet clear of obstacles to the search for a mate. Now and again the courtship hunt would even be illuminated by the headlights of helpful drivers. Actually not so helpful: we found it easy to miss most of the 'manders with a little care—"We brake for newts" came to mind as a suitable sticker—if not the moving targets of frogs. Yet, sadly, the roadway lay littered with flatter-than-usual salamanders. So the adoption of the public right-of-way as a rendezvous proves maladaptive after all.

In particular, we noted the massive carcasses of northwestern salamanders. Two of these, both nipped clinically by the head and otherwise intact, we collected for examination at home. The postmortem proved fascinating. John McPhee, in his essay "Travels in Georgia," has written of the pleasures of road-kill cookery. I find that D.O.R. (dead-on-road) animals, if fresh and intact, furnish elegant opportunities for highly instructive dissection. Thus may the lessons of college morphology classes continue without having to kill the subjects, and thus may poor road-kills be appreciated, even by those disinclined to consume them.

They turned out to be a male and a female, the former measuring 21.5 cm and weighing 30 gms; the latter, 24 cm and 60 gm. That made the female over nine and one-half inches and two ounces, the size of many a respectable mammal, a truly imposing animal. Though just an inch longer than the male, the female was twice his weight. The difference seemed to be accounted for within by elaborate, convoluted, paired masses of white tubes, stretching to more than 50 cm. I took these for ovaries. The actual eggs, consisting of seventy-five or so in each of two jade-green clusters, lay in yellow sacs at the ends of the oviducts. What surprised me was that the other salamander, while possessing the swollen cloaca of the male and apparent testes, also had the ovaries—though much smaller and less well developed and with no sign of eggs.

Perhaps I should not have been surprised after all, since hermaphroditism is well known among amphibians. In *The Sex Life of the Animals*, Herbert Wendt described how the sex of many young frogs is indeterminate at first. Among obvious males and females, one finds numerous examples of AC/DC adolescents bearing juvenile gonads of both sexes. Apparently, these may eventually go either way. Most of the hermaphrodites will mature into males, some into females. Salamanders are not mentioned, but my sample of two road-kills—one definitely female, one ambiguous—seems to show that these creatures too may be sexually ambivalent in youth.

I suspect that if I knew all the secrets of salamander sex, inside and out, I should have a capital tale. How, for example, do they locate one another in the dark nights of their roamings? It isn't likely that the shine of their liquid eyes and glistening skin in the dangerous headlights helps at all. The scent of pheromones released in the waters of their destination probably attracts salamanders to one another, at least those that have made up their minds and bodies about which sex they intend to be. Males release their sperms in a gelatinous spermatophore at the proper time as determined by sometimes elaborate courtship procedures; the females then collect this gift with their cloacas. I'm sure we lack many facts, among them the nature of satisfactions gained, if any, through such a chaste form of fertilization. One assumes that, the act completed, hormones allow the mating frenzy to subside, whether the union involves copulo or not; and that must be a form of satisfaction. How little we really know of different strokes.

One thing we do know is something of the way in which these big amphibians repel predators (as they must, for they offer such easy prey in their precourtship perambulations on land). These rubbery animals possess thick, spongy glands behind their eyes, along their sides, and up and down their muskratlike flattened tails. When I squeezed these with the dissecting needle, they fairly oozed a toxic fluid much like milkweed juice. The thick, white stuff stood by in such ready quantity that, from one pressed gland, it squirted into my face, several inches away. I hurried to wash it off. My fingertips were roughened and tender from handling the copious, milky venom. Who would eat such a beast?

Our rough-skinned newts have their venom too, which they advertise with their bright red bellies—a form of self-promotion known as warning coloration. Many brightly colored, distasteful insects, like the monarch butterfly, do this. But for the population to acquire protection, the potential predators have to learn to avoid them through initial unpleasant contact. Hence, the famous photograph of a barfing blue jay in Lincoln Brower's 1969 *Scientific American* article on ecological chemistry. The naive jay, having tried one monarch and found it extremely unpleasant, has been educated: it represents no further threat to monarchs. Look-alike viceroys, though more palatable, gain protection by mimicking monarchs. A news item in an Oregon paper a couple of years ago suggested that at least one logger had less sense than a blue jay. Dared to drink a newt in his beer, he did so—and died three hours later from acute toxic reaction. No more red-bellies in the beer in that tavern, I'll wager. The ploy worked for the local population of newts if not for the test case.

Another western amphibian known for its venomous glands, the northern toad (of the lovely name *Bufo boreas*), should by all reckoning proliferate throughout the moist, green land of Willapa. But, as we know from Thornton Burgess and Kenneth Grahame, old Mr. Toad is a willful animal. Both writers were as much sharp watchers as they were clever storytellers, and they knew an animal's nature before investing it with character. As Rat and Mole found, Toad just can't be trusted to do what you expect him to do. And so it is here. I can't find a toad for the life of me, nor discern any limiting factors for their living here. Surely not the Rot Factor that I believe discourages butterflies, for they thrive in the moldiest spots. Can't be cold, for I've met them high in the Colorado alpine. Yet they are just not about as they are elsewhere. I have never seen a toad at Swede Park, apparently as fine a Toad Hall as ever there was; and I know of only one or two places where toads have been spotted in the Willapa Hills by reliable observers. The fact that from time to time they get trapped in the ponds at the Naselle Salmon Hatchery proves that they do in fact occur in the area; but why so sparse or retiring? I wonder.

Frogs, on the other hand, abound. Of course, frogs tend to be even more moisture-loving than toads, so their predominance here should not be too surprising. Both Pacific tree frogs and red-legged frogs (more euphonious names: *Hyla regilla*, *Rana aurora*) appear in the damp recesses of the woods in great numbers. Both take to water, of course, to breed. The spring-green tree frogs begin calling in the rushy meadow-swamps in February or March and carry on their nocturnal disquisition for months. It is always the same, each evening of each spring: first a single, tentative "rigit," becoming hortatory as more voices arise, finally a full-blown shout as inflated trills join in concert; then tailing off in exhaustion, ennui, or simple satiety. But before the croakers rest, each a jade and jaded voice box with a thousand-watt amplifier, they've given background music to an entire season.

Red-legs issue their quiet croaks from ponds in the woods, often calling underwater if the books are to be believed. Later in the year they show up in the forest. Some, almost as large as modest bullfrogs, startle the prowler of the fern banks by stirring far from any water. How, in a dry summer such as this just passed, do they remain moist? By the dry time the tree frogs too have shifted position. When the water withdraws from the ponds and the rushy bottoms, they take to the trees and bushes. And by August or so, all we hear is the occasional katydidlike croak from an oak, or the feeble "braack" of a leaf-mold lurker. The ponds and river backwaters boil with polliwogs, and soon the tiniest frogs you ever saw—bright green mites with raccoon masks and golden eyes or mottled brown hoppers with strong pink thighs—populate the woods to wait for winter wet and yet another spring.

Bullfrogs, introduced, float in shallow lakes with their huge eyes poking out: joke frogs. Other native species we have yet to find in their favored waters. One of these species recorded for the hills has shown up only as two tadpoles. Ed Maxwell, watchful keeper of the Naselle Salmon Hatchery, who spotted most of the local toads I know of, also found a pair of immature tailed frogs, one each in the Naselle and Gray's rivers.

The tailed frog (*Ascaphus truei*) has to be regarded as one of our most interesting organisms. Its name originates in the caudal flap that covers the genitalia of the male. This "tail" is actually a copulatory organ that facilitates a kind of internal fertilization—a sexual practice engaged in by no other frogs. Such an adaptation probably arose to permit mating in the swift streams occupied by the species. Without it, the spermatophore would be swept away on the current more often than not.

The tailed frog's distribution is as interesting as its sexual structure. Restricted to the Pacific Northwest *sensu lato*, it occurs in many separate streamside populations in California, Oregon, Idaho, Montana, Washington, and British Columbia. Yet its only close relatives are three species of terrestrial frogs in New Zealand. Lacking the genital "tail," these frogs nonetheless share sufficient traits with *Ascaphus* to make the relationship clear.

So how does a frog-family founder get from New Zealand to Oregon, or vice versa, and why doesn't it show up elsewhere in between? Consulting my old college notes from the late Professor Frank Richardson's excellent class in zoogeography, I find that these long-lost cousins are prime examples of relicts. (Relicts are species left behind on outposts when their formerly more widespread kin become extinct in between.) I read, too, that the tailed frog and its relatives are considered to be the oldest, most primitive family of frogs. So presumably they evolved long ago, whether in the South Pacific, the Pacific Northwest, or somewhere entirely different; dispersed; then died out in all but the two regions.

Frogs get around, like other animals, either under their own steam or by rafting across seas in storm-tossed bundles of wrack. Continental drift may have affected their earlier whereabouts. They drop out in the intervals due to all the catastrophic or gradual changes that bring about extinctions everywhere. The resulting pattern illustrates disjuncture, whereby once-widespread species have withdrawn over a long period to isolated fragments of their former range. Disjunct relicts are not infrequent in the floras and faunas, but few examples are quite as dramatic as the tailed frog and its New Zealand relatives.

Having failed to find it here for years, I suspected that isolation might have excluded the tailed frog from the Willapa Hills; or else, the fact that it seems to favor cold-water streams, whereas most of ours are relatively warm due to low altitudes and lack of glaciers or snowmelt upstream. But Ed's discovery would seem to prove that the animal occurs in Willapa. Even so, he might have missed it but for another remarkable trait of the tailed frog: an anchor against the current.

The tadpole's mouth is as interesting as the opposite end of the adult. Possessed of a sucker, it clings to stones in the rapid streams where the species lives. When Ed worked at the Klickitat Hatchery in the southern Cascades, he noticed large numbers of tailed-frog larvae (his coworkers thought they were leeches) attached to the cement side of fish ponds, where they apparently fed on algae and microorganisms. That experience keyed Ed to watch for the unique polliwogs when he moved to other hatcheries. After he'd found the first one in the Gray's River, we figured it just might have been an introduction, since Klickitat fish had been moved to Gray's River, and the tadpoles could have hitched along. But Ed's duplicate discovery in the Naselle River clinches the case: tailed frogs do dwell in the Willapa Hills.

Still, in order to find the adults and finally see the tail of the male, we shall, I suppose, be obliged to spend still more time alongside the rivers of these hills, perhaps at night, watching for the eyeshine of the nocturnal frogs, or overturning stones by day. No penalty this, for we love to do such things. Finding the tailed frog, which I have never seen, would be a great satisfaction. And in such pursuits, the pleasure lies largely in the search itself.

I remember a day spent in Mount Rainier National Park with my friend Noble Proctor, a New England naturalist intimately associated with the Roger Tory Peterson Institute. Noble had never seen *Ascaphus* either, and his blood was up for finding it. Avidly, we began turning cobbles in the Nisqually River. Eventually we were forced to give up as our hands stiffened in the glacial meltwater and the day's light ducked out. But the search added one more shining coin to our joint treasury of Mount Rainier memories, and we know we shall try again.

Meanwhile, my wife, Thea, and I intend to join the Maxwells in seeking tailed frogs in the rivers of these hills. I can hear the tavern-talk of more conventional hunters, bringing back their elk or deer, declaiming upon the sight they saw: ". . . and here were these people hunting some goddamned, fancy-pants frogs!" To each his own.

Superbly adapted to life in swift streams, the tailed frog seldom strays from that habitat. Different amphibians favor quiet waters. The newts are among these. When not actually immersed, newts suit themselves to wet places generally. Not only do they wander over the green sponge like true habitués, but they also display such exuberant fecundity in their season that they can only be thought of as "happy" with their surroundings—in an ecological sense, of course, unless you will grant me leave to personify my newts after the fashion of Burgess and Grahame.

When I took a group of local schoolchildren on a nature walk to a nearby lake last spring, the prolific feeding newts stole the show. But only later, when ponds diminish into puddles, does the real spectacle of massing 'manders exert itself. In the mossy bottoms of Hendrickson

Canyon, a nearby old-growth forest remnant we visit often, we found perhaps half a hundred rough-skinned newts crowded into a pond the size of a backyard wading pool.

That was nothing, I guess, compared to the spectacle observed by Ed and Cathy Maxwell this past hot summer in a small swamp fringing Elochoman Lake. Mostly dried out, the swamp offered only puddles for exploration. Crossing one of these on a log, they gasped at the sight of a seething clot of newts some three or four feet across and several inches thick, writhing just below the surface.

Ed, a fisheries biologist accustomed to estimating numbers of small, squirming animals in the water, reckoned there must have been a thousand or more newts taking part. Every one, it seemed, constantly sought the center of the mob. The attraction remained obscure—whether the shade of their own bodies, a concentrated food source, the pheromones of a receptive female, or some other sort of chemical triggered the mass assembly, we can for now only guess. Feeding frenzies take place among sharks and great mating orgies among certain toads; but to what end thronging newts?

All this talk of newts and the Maxwells brings to mind a favorite anecdote. In his charming classic *Ring of Bright Water*, Gavin Maxwell tells of his walks through the streets of Kensington with his Indian otter, Mij, on a leash. This took place during a brief, necessary stopover on the way to the better otter lodgings of the author's Scottish retreat, Camusfearna. Not too surprisingly, the Londoners encountered on these walks often failed to recognize Mij as an otter. The range of their guesses ran the gamut of mustelids, from mink and weasels through ferrets and badgers, and included beaver, seal, walrus, and brontosaurus. (I experience the same range of guesses as to the identity of an otter on a ring I wear.) But the strangest and funniest speculation came from a navvy who figured Mij for some sort of a great *newt*!

Actually, that London laborer, perhaps familiar with the smooth newts in his garden pond, showed a rare perceptivity on a gestalt, or

Aristotelian, level. Or so it seemed as I observed newts and otters in propinquity last spring. At a small lake near Olympia, a beaver-dammed portion of McLane Creek, I had the good fortune to come upon a fishing river otter. Again and again it dove, snatched a shiner, rose, and crunched it in its sharp, strong teeth, while apparently standing on its tail in the shallow water. Eventually the otter hauled out, basked, and preened on a log for a spell, then resumed its fishing. Sated at last, it dove once more, came up in the center of a hollow, upright cedar stump, and bedded down on the top among huckleberry shoots and soft, rotting cedar fibers. I was privileged to watch a wild otter asleep, unguarded and thoroughly at rest. This was a deeply moving experience for me. Then passersby on the pondside nature trail disturbed a beaver, causing it to slap its tail on the water. That noise alarmed the otter and caused it to bolt down into its cedar holt.

The next time I went to McLane Creek I watched newts at the same time as the otter. From my vantage on a wooden fishing/bird-watching platform, I could see the otter clearly; could watch the bloody-shouldered blackbird clamber up the cattails before issuing its screeching song, and the newts hunting small aquatic game at my feet. What struck me so was the remarkable similarity in form and function between the newts and the otter. In feet, in tail, in sinuous trunk and pellet-head, in buoyancy and fluency, in gentle glide and rapid stroke—here were two unrelated animals performing much the same act with much the same movements and appearance.

We are dealing here with convergent evolution. If something works well, it will likely have evolved more than once. Species possessing analogous features may give the appearance of close relationship, or at least show striking similarity of form. Coincidence has nothing to do with it: take animals of diverse descent, add water, mix with natural selection, and you get newts, and others that look not a little like newts.

Some of the others, besides otters, include muskrats, beavers, and the introduced nutria, or coypu. Each of these aquatic mammals occurs in the Gray's River valley. Otters we sometimes spot midstream while out in the canoe, but never yet behaving as brazenly as at McLane Creek. Beavers show themselves chiefly in their work—felled willows, great cottonwoods chiseled halfway through, then abandoned. Around here they live in bank burrows, and you see them chugging past like ponderous, submerged locomotives at river's edge. Up at the Lake of the Newts, they even fell hemlocks (seeming not to mind getting the acrid pitch in their vibrissae) and build the standard dams and lodges and ponds and sluiceways that so reorder the countryside.

Muskrats show chiefly in the little Lily Pond down valley that gave the first dairy here its name. A tenth of the mass of beavers, they steam around among the lilies like Little Toot to the beavers' freighters or sit preening on emergent logs.

Now we wait to see whether the muskrats will survive in the Lily Pond, for the nutria have moved in, and certain texts advise that the latter are prone to drive out the former. Nutria came from South America as a potential fur source. As with most alien animals (George Laycock's potent term), damage is done in the introduction. The big, heavy rodents burrow into banks and graze the uplands, causing drainage leaks and loss of forage. They compete with native furbearers for their primary food, a range of aquatic plants. And in a manner not entirely clear to me, they are supposed to exclude native aquatic mammals. We will watch the Lily Pond and see what happens with the muskrats.

In any case, the nutria make good watching material. Unafraid, they permit close approach as they haul their great, squared bulks over the pasture banks, their huge orange teeth working the grass like a mowing machine. Wombatlike but generally black, they add a definite presence to an already-altered ecosystem, and they are somewhat more interesting than cows.

When any of the aquatic mammals take to shore to preen themselves, you can see how fur has become a waterproof fabric. The thick guard hairs separate into silvery, spiky locks that adhere so closely, and yet hold so much air, that they must let very little moisture through to the skin. Indeed, as the soft underfur lies exposed to the kneading teeth or claws of the animal or its mutual groomer, it looks as dry as could be. Immersed in a northern river, we would quickly become soaked to the skin and chilled to the bone. But not these creatures. The accomplishment of adaptations such as body fat, fine, downy close-fur, guard fur, or feathers, and special oils is to permit warm-blooded life in cold water.

Therein lies much of the devastating effect of oil spills upon such animals. By ruining the necessary nap and natural oils of the bird or mammal, petroleum products destroy their water repellency and admit cold water to the skin. Or if the oil itself doesn't do it, detergents and other cleaning agents may. Further, the action of the currents and tides tends to carry the oil deep into coves and channels and inlets frequented by aquatic furbearers, exacerbating the likelihood of injury—this was demonstrated by the disastrous *Mobiloil* spill in the Columbia River two years ago.

As long as the rivers remain pure, the holts unflooded, the lodges undynamited, and the animals themselves untrapped (the perils of the aquatic life are many), these animals will thrive in watery regions. Of course, no animal accustomed to life underwater is going to mind a little rain.

For terrestrial mammals, it's a different matter. Unequipped for immersion, how do rodents and rabbits, ungulates and insectivores, bats and bears and bulls, keep from drowning in the deluge or, at least, from becoming discouraged to the point of emigration? And as for the cattle: why don't they simply rot out there, I often wonder from within my warm, dry cell. (My friend Marilyn Gudmundsen, wife of a beef farmer, told me recently that they *do* rot: at least their hooves can rot in a long, wet winter when the mud never dries for month after month.)

Some mammals just don't take well to the rain and do not occur here. Neither foxes nor rabbits do well in Willapa, though the surrounding ranges have plenty of each; have they a limit to their tolerance? Hares occur but sparsely—a rare condition for their ilk. Skunks and porcupines maintain lower numbers than one might find in most drier parts of the state. I suppose this could be an illusion due to the dense vegetation, making them harder to spot; but their sign and the number of roadkills and direct sightings all tell me they are not common. Maybe it is something else about the Willapa Hills that keeps down the numbers of mourning cloaks, toads, skunks, and hares, but the obvious scapegoat is the rain—as for everything else.

Raccoons too seem less abundant in the Willapa Hills than in many another locale. Clearly they love water, but just how much? They take to hollows or leafy crotches for the storms, faring at least as well as John Muir did climbing trees at such times. The plentiful opossums, whose hobby seems to be patching the roads with their own hides, just get sodden in the rain and look that way, with their matted and sparse gray hair overlying dense, cottony, skintight wool. If they are half as miserable as they look on a rainy night, it might help to explain their terrific mortality on the roadway. The human suicide rate is known to be especially high among transplanted southerners in the rainbelt of the Northwest. But sunbelters, human or possum, seem to increase here nonetheless. The accents in the country bars attest to it. Arkies brought the possums in the first place, during the Depression, goes one version; and they've brought their coonhounds too, perhaps in part explaining the dearth of raccoons in some parts of the countryside, whereas they proliferate in the middle of Seattle.

Opossums, raccoons, and coyotes all do well in human habitats, provided an adequate amount of cover remains. This mixed kind of countryside just suits coyotes. They can hide out in the rough tangle of the second-growth woods, then range into the valleys at night after rodents, afterbirths, and cats. They seem to have come to terms with the precipitation—or might their coloratura yips and yowls be as much bitching at the rain as celebrating a clear, moony night?

The prey of coyotes, the burrowers and hole-dwellers, seem to exist to feed the predators of the world and can have little security. My cat will persist in catching, for example, the beautiful tawny and chestnut, long-tailed jumping mice called *Zapus*. And smaller carnivores step in when the larger ones fail: every dead chipmunk I've seen this summer has borne botfly larvae as large as its own ears, eating directly into its gut. (Perhaps that's why I came to find them: the cat, maybe as disgusted as we, ignored them after a trial pounce.) Still, when they do find rest in between the constant chase and scratch, these small ones presumably make themselves cozy in quarters lined with mosses, lichens, and thistledown. The very vocal, none-too-common Douglas squirrels or chickarees have their tree holes, Townsend's chipmunks their endless woodpiles. Cold comforts these, perhaps, but at least a roof from the rain.

But what of the moles? How do they keep from drowning like rats in their subterranean tunnels? Perhaps they do drown, in great numbers, and if they didn't our gardens and fields would be completely mounded and undermined by mole-workings instead of only mostly so. Moles make two kinds of burrows, shallow for feeding, deep for breeding; both, it seems to me, would be subject to inundation. In fact, whenever the floods rise in our valley and the water table forces itself out ready-made geysers in our garden and cellar, one of the most insistent spouts always issues from a certain mole hole.

A particularly compelling chapter in *Wind in the Willows* deals with the theme of home and the gentle comforts of same. To convey this, Kenneth Grahame employs Mole and his return to his subterranean dwelling after a long absence. I admit that I really have to work at believing a mole's home might be so cozy. It seems to me that moles, shrews, shrew-moles, garter snakes, and the other users of holes must be the most vulnerable of all to the rain, more so even than the deer and the cattle that must abide right out in the open. They, at least, are not likely to drown outright in any but the most severe deluge. These burrowers must all have some way of dealing with the draining of the rain. Whatever it is they surely make it, for each year when the rains back off, spring brings new evidence of their presence in the form of newly tilled mole-mounds, cats' trophies, owl pellets, and occasional live appearances.

Equally evident is the fact that at least one burrower does very well indeed among the damp humors of the Pacific Northwest. In fact, it occurs nowhere else: although its ancestors lived across the Northern Hemisphere in the Pliocene, and members of the same family occupied most of North America in the Eocene, it has since been restricted to a single West Coast species. Occurring roughly from San Francisco to Vancouver, and between the sea and the high peaks, *Aplodontia rufa* is considered the most primitive of living rodents.

Commonly known as the mountain beaver or aplodontia, it has also been called the sewellel (an Indian word for a robe made from its hide). The name mountain beaver is unfortunate; people tend to confuse it with beavers, muskrats, and coypu and to assume that it lives in the water or in the mountains. In fact, aplodontia occurs in moist, often streamside conditions in the lowlands and foothills, where it can burrow readily into the soft, duffy soil. It has been known to climb trees rarely, and for all I know perhaps it can swim; but it spends no time in the water proper, according to the literature. About muskrat-sized and -shaped but lacking any apparent tail until you look very closely, the sewellel (I like that name best) shows itself infrequently.

My own encounters have been restricted to two or three nocturnal spottings, when the reddish pellet pattering over the wet road resolved itself in the headlights into aplodontia; and a single road-kill found near home, so pancaked by repeated poundings that it took me two days to figure out what it was. Stretched by truck after truck into a big brown tube, it looked like a very fat weasel's remains; finally the minuscule tail gave it away.

Yet, though we see the sewellel but rarely, its evidences abound around us. Like Townsend's snail and the white trillium, aplodontia is a classic creature of Northwest ravines. No one can pad through the sword ferns in search of chanterelles without eventually caving into a mountain beaver's burrow. Up to ten inches in diameter and like a winterknot of snakes in design, these tunnels put up a formidable barrier (perhaps second only to devil's club) to anyone wishing to navigate these ravines.

Sewellels make their presence known in another way as well. I recall a day afield with Melody Allen, Portland conservationist, in an Oregon old-growth forest slated for logging if a deal could not be hatched by The Nature Conservancy, for whom we both worked. We marveled properly at the big Douglas-firs and admired mossy knolls, potted with blossoming saxifrages. But what struck us most that May day in the Cascades foothills was the haystack of the sewellel we came upon. There, in an animal runway between the giant cedars, we found it: as neat an array of cut flowers and greens as you could imagine. All lined up, scissor-clipped ends altogether, certainly the work of some sylvan gardener, florist of Faerie. And so, I suppose, it was; only the hand-maidens to Titania were furry, primitive rodents, rather than fairies in gossamer.

But really, is it any wonder that tales of little folk should arise in the greenwoods of the world, given evidences such as these, and people's proclivity for imagining and embroidering rather than biding by night to observe and explain? The young ferns, saxifrages, oxalis, inside-out and piggy-back flowers, grasses, and bedstraws that made up this hoard looked a perfect fairy-queen's bower and selected for looks rather than nutritive value. Perhaps the sewellel is a meticulous animal, more than the pika of the high country, whose vernal haymaking must be hasty in preparation for winter's rigors. Pika stacks look more like broken hay bales than floral tributes. Aplodontia, on the other hand, occupies mild territories where, even in winter, something green to eat may always be found in its ravines.

None of this speaks to the ancient animal's ability to withstand wet. Make no mistake—its habitats are among the wettest around, even in Washingtonian terms. The water off the tops funnels down the ravines like recycled beer in the bog of a British public house. The rodents tend to stick to the slopes; you seldom find them in the very bottoms. They burrow in deep duff of needles and decomposing moss and rotting wood, young soil that drains well, among supporting roots of trees. So I should think the chances of their burrows flooding far less than those of the moles of the valley floors, who must grow gills in winter. (The mud-puppy stage of moles is another of the evolutionary enigmas missed by the phylogenists, like the evolution of otters from newts.) I am in no position to know whether mountain beavers enjoy more than the coldest, dampest comforts in their lairs, nor whether this most primitive of rodents possesses "enjoyment" in its repertory of sensation. But one feels that fifty million years is too long to spend uncomfortably. Roughly the same sort of creature for at least that long, aplodontia shows a degree of evolutionary complacency that must say something about contentment.

Evolving along with *A. rufa* have been its parasites, for all creatures have them. These attest to the species' antiquity, for they are both unique and specific to their host. They include both the largest flea in North America and some of the few parasitic beetles known, the platopsyliids. Even whales and salmon, in their fully aquatic existence, bear external parasites; so we should not be surprised that any damp-land animal would do so. And there is no getting away from them, whether you roam the seas or remain within a few hundred yards of your birthplace, as the sewellel is thought to do. It is stuck with its big flea and bloodsucking beetle just as, being endemic to the Northwest, it is stuck with the rain and makes the best of it.

Swallows too have their fleas, passed from parent to hatchling in the nest. But at least they can leave the rainiest months behind, and do. Gilbert White was unwilling to admit the total desertion of Selborne by his favorite birds in winter and kept on seeking their hibernaculi all his life. Now the evidence is in: they do, indeed, leave, fair-weather friends they.

Our birds of passage cope with the wettest season by simply going elsewhere. But this may have little to do with rainfall. Just as birds forsake cold-weather places for warm in order to find food rather than to avoid snow and ice, our departers probably do so because of the seasonal decline of their necessary resources. Whether they would migrate if adequate food were available, one can only guess. But those birds whose food supplies remain available throughout the year (flickers, juncoes, towhees, and crows, for example) stay home—and handle the hose spray off the sea and out of the clouds in their own specific ways.

As with aquatic mammals, many birds like nothing better than a watery world and seem (in our minds, at least) to revel in the rain. Ducks, of course; their numbers swell with the margins of Larson's pond in winter. The kingfishers, whose rattles sew up the space between the covered bridge and the trees upstream; the great blue heron, crusty croaker when disturbed in the dusk; the bald eagles, and the ospreys, and the rafts of mergansers—all these fishers rely on the water to get them through the year and take what the water dishes out in return. Others just abide (dispirited, damp fluff ball of chickadee; crestfallen, faded-blue jay), and, like the snakes, few lizards, and fewer humans of Willapa, make the most of the basking sun when it comes.

I watch the birds a lot and have many sights to share. They will have to wait for another book, along with the moths, and the mushrooms, and the fish. But there is one vision I must include here, for it has to do with abiding in the rain and making the most of the sun. It happened in September 1985, as Thea and I drove eastward along the Columbia River. Just before the Wahkiakum-Cowlitz County line, I spotted two large birds spread-eagled in a snag, stopped, and went back. Actually, there were five big birds in the snag, and they were spread-vultured. The odd bird stood on the very top of the tall, dead tree; beneath it, two more turkey vultures perched side-by-side on spindly branches alongside the trunk and, beneath them, another pair. Each bird held its six-foot wings outstretched entirely, much more erect and rigid than cormorants in a similar posture. As cormorants do, the vultures were drying out their wings and absorbing the warmth of the sun through their black feather cloaks. It had rained for days and nights and days, and I can imagine the scavenging great birds were thoroughly sodden, as well as totally fed up.

Arrayed as they were, symmetrical and steaming in the early beams, the pyramid of vultures resembled ornaments for some macabre anti-Christmas tree. All gray and black, no greenery, no color at all save the birds' lurid red faces, this pastiche could have seemed Stygian and bleak. But all that is flimsy fantasy. Far from sinister, this magnificent fivesome displayed all the bright cheer of those who feel the sunshine on their backs after chilling to the rain on their heads for a long, cold time. In the spread of their great wings they embraced the sun for us all. And that's fancy too, but I prefer such a view.

The rain fell. The wet birds huddled. The sun came. They spread their wings and grew warm and dry.

Before we left, the clouds closed, and one by one, the vultures folded their wings again, reluctantly, it seemed. Last of all, the bird in the angel's position atop the tree drew in its broad black vanes. Then the rain began to fall.

# Part III

## Hands on the Land

# The Sack of the Woods

The very fact that I wrote the letter shows my callowness when I came here. I must have thought they were honorable men, these timber bosses. The loggers I knew were not only honorable but likable. Tutored at two forestry schools in the romance of the early logging days and the righteousness of modern industrial sustained-yield forestry and determined not to let my environmentalist bias get in the way of good relations with my neighbors, I somehow expected more of the company. There is that in goodwill that expects reciprocity. When it doesn't come, when you extend your hand and get kicked in the face—that's when you wonder.

The Timbered Tor stood across the valley from Swede Park. It was just a second-growth hillock at the base of a bigger Willapa Hill. The south bound of the Gray's River valley is a broad, flat-topped, old river terrace backed by emergent lumps of old ocean floor. The part to the west had been clearcut in the mid-seventies and stood mostly denuded and brown. The part to the east had been left for longer and never sprayed, so now it formed an almost New England-like mosaic of hardwoods and softwoods, pleasing to the eye. And between these lay the Timbered Tor—only ten or fifteen acres or so but the largest tract of tall, old, second-growth Douglas-fir, Sitka spruce, and western hemlock visible from the valley floor—the centerpiece of the scene.

Fresh from England where the range of topographic terms equals that of the playfield of proper names, I called the logged-off mountain the Grim Fell (for it was very grim); the wooded, unmanaged hill already had a good name, Elk Mountain; and the tall-tree knoll in between became the Timbered Tor. A fell, in British geography, is a barren hill; a tor is a jutting hill.

From Swede Park, the Timbered Tor dominated the view along with the covered bridge, the river itself, and the water meadows. Through my window each morning I watched the mists disperse among the green-black ragged tops of the trees upon the tump of the Timbered Tor. I came to love those trees more and more. In a land where trees fall daily, one can't afford to fall in love with any particular stand. But I did and began to worry about the fate of this beautiful grove. The more time I spent watching them, walking among them, studying their community, the more I wanted these trees to stand and worried that they might not. My neighbor Veryl Chamberlain used to own the land on which they stood, but he told me that Crown Zellerbach had bought it. He didn't know the company's plans.

So one day, unable to sit by and enjoy the Timbered Tor without some word of its future, I wrote the letter.

19 October 1979

District Manager
Crown Zellerbach Corporation
Cathlamet, wa 98612

Dear Sir:
I am a local resident with an interest in a parcel of land which I understand belongs to Crown Z. I would very much appreciate your taking time to respond to my query.

The land I speak of is a forested stand, second- or maybe third-growth, situated in the sw quarter of Section 17, T10N R7W.

This is just opposite (south) of the historic covered bridge in Gray's River. It is the only mature stand of timber in its vicinity, and I would guess the stand occupies no more than twenty acres. I believe this land was purchased relatively recently by the company. I have heard that it could be logged off soon.

My interest in this parcel comes from two or three different angles. First, my wife and I live directly opposite the timbered hill, and it makes up an important part of our view. These are the only big trees visible in the same view as the covered bridge, the hills around there having been logged off recently. Second, I am a writer and a biologist, and I am interested in doing some long-term research on the hill. I checked it out when we arrived here, not knowing to whom it belonged, and found that it is quite rich in wildlife and plant life. I don't know whether anything rare or unusual lives there, but it is certainly a rich site, the best in our vicinity by far, as most of the rest is under intensive management. My wife is an artist and she does wildflower prints, often finding her material on the land in question. As I mentioned, I would very much like to be able to carry out long-term research and writing on the natural history of that forest.

Obviously, we would like to see the area set aside from logging. We appreciate the vital role of the timber industry in this area and we certainly don't want to impede that. But this is a very small stand, and I wonder if it is necessary that it be cut. I would be interested in talking about options for this tract, which might involve purchase by myself or a conservation group, or donation of a conservation easement on the land to a nonprofit group such as The Nature Conservancy. There could be significant tax gains and public-relations advantages for Crown in such a case. I have worked in this area of private land conservation and I would be very happy to have an opportunity to talk with you or any of your colleagues about the possibilities.

> In any case, I would appreciate being informed as to whether you have scheduled timber operations for this site, and if so, whether they might be delayed long enough for us to mutually explore any alternatives.
>
> Please feel free to copy my letter to any other Crown people who might be interested or willing to look into this matter with you. I will look forward to your reply, when convenient.

I received no reply. After a few months I decided I never would, that such a tiny parcel of land couldn't interest the corporate giant anyway. And I proceeded to try to forget about it. Then one morning I awoke to the sound of howling chain saws, coming from right across the valley.

"They're cutting the Timbered Tor!" I exclaimed without looking as I jackknifed upright in bed. I was right: scanning the far side of the valley, mistless that day, I could see the loggers moving antlike among the hundred-foot trees they were about to fell. I could see the orange trucks, and the bulldozer scratching a road diagonally across the slope. I could hear the sharp, irritating hoots of the signal whistle. The big trees began to tumble.

It only took a few days for them to finish it off. They left nothing, even took some trees right on the unclear border with the neighbor's land. Mud and sawdust, slivers and stumps, remained where the beloved wildflower wood had stood.

At first, bitter at the fact and resentful of the company's failure to even discuss it beforehand, I consigned it to one of the mounting acts of philistinism one encounters in the country; acts, I thought, that cumulatively would one day cause me to avert my eyes and go. How could they trash that fine little forest for the few trees it held, I wondered, without even discussing the matter? But I noticed that many handkerchief patches of woods were going, as Crown liquidated their mature holdings prior to their rumored pullout: cut and get out, just as in the bad old days. It appeared that it was quick profit, and quick profit alone, that motivated the managers.

But it was worse than that. When I told the tale later to a friend who knows the woods in ways I never will, he laughed, with a disgusted snort. I'm not sure whether his contempt was aimed at me or the company as he said, "Why, you just performed a timber cruise for them—they oughtta pay you for that! They probably never woulda noticed that little patch of logs if you hadn't rubbed their noses in it."

I flushed as the likely reality dawned on me: *I had fingered the Timbered Tor.* I might as well have let the contract on the cutting and called for bids in the *Eagle.* Loggers have a term for the likes of me: they call us tree huggers. True, I had hugged some of those old firs and hemlocks; but now I'd given them the Judas kiss. I saw to the logging of the trees I loved.

Of course, I cooked in chagrin and self-recrimination over this for a long time. Every time I looked out at the raw, miniature clearcut, the view served as a reproach instead of the balm it had been. I still regret my impatient advance, although I realize that they might have cut the trees at some point anyway. But most of all I marvel at my own ignorance: I really thought they would answer that letter.

Not long after the felling of the Timbered Tor came the poisoning of the Grim Fell. It is common practice in the Pacific Northwest for logged-over lands to be sprayed from the air with herbicides designed to prevent the regrowth of alder and other hardwoods and brush. This permits the desired conifers to take over faster. The use of herbicides on public land has received a great deal of attention in recent years due to well-established health risks. Court cases brought in Oregon resulted in the banning of the dioxin-bearing chemical 2,4,5-T, related to Agent Orange, on both public and private forestlands. Another decision placed a moratorium on the use of the related chemical 2,4-D on the public domain until a proper environmental impact study could be conducted.

In another fit of naïveté, I thought these actions might cause the private companies to think twice about spraying their lands. Not a bit of it. By their own admission in public statements to the press and to anyone who asked, they intended to continue using 2,4-D to suppress alder and scrub on their timberlands. And so they did. On May 10, 1983, a helicopter sprayed the alders on the Grim Fell, back and forth, for a long time. What an awful time to spray, I thought, just when the alders and shrubs would be full of birds' nests. I watched ruefully as the deathcopter blurted its venom over the greening scene.

More philistinism. I tried to console myself with the words of E. B. White in defense of his imperfect but beloved Northeast. In reply to Bernard De Voto's description of Highway 1 from New York to Maine as a "slum," White wrote: "Familiarity is the thing—the sense of belonging. It grants exemption from all evil, all shabbiness." I tried to apply White's logic. But with the razing of barns, vandalism on the covered bridge, the butchering of K-M Mountain and the Timbered Tor, and now the spraying of the river terrace hills, I wondered how much a sense of belonging could be worth. Would familiarity soften these repeated blows?

Those episodes, though painful, were instructive. To live here, one must understand something of the way the woods work—and I mean economically, not ecologically. For the great majority of the small number of people here, the woods have meant money: wages for workers, profits for bosses, and coins on the counters of the shops and taverns; kids for the schools, taxes for the roads, and customers for the utilities. That's the way it's been since the felling of the giants. Farming and fishing have filled in around the edges and sometimes taken over here and there for a while, but logging has been the economic mainstay of most of western Washington and Oregon since soon after they were settled. Understanding this, one can hardly begrudge the taking down of trees, and in that context my concern over the Timbered Tor may seem petty and unrealistic. It's a logging world, and you like it or lump it.

To a point. But it turns out there is more to it than that. My complaint has to do with the way this particular episode was carried out and with the way a lot of logging has been carried out from the start. It is only starting to come home now that more and more logging-dependent people are lumping it instead of liking it. And in that respect, the sack of the Timbered Tor and the sack of the woods are just mini- and jumbo-sized versions of the same sad tale.

I almost despaired of trying to write this essay. The woods of Willapa have been ravaged, along with its soils, rivers, and communities. It's a simple tale in many ways—great trees gone, boomtowns busted, fragments of forests struggling toward a kind of recovery, only to be logged again, too much, too fast. But the telling of it cannot be done simply. It carries too much satisfaction and pain, injury and age; it holds too many promises and lies; it weighs with too much labor and hope, profit and loss, heroism and hypocrisy.

In the end, it's just people and trees, money and time. And it's living with what we have wrought, coping with how the ax has fallen. But no writing of it can tell the whole tale. All I can do is to give my view, which will no doubt differ from the telling others might give it.

This is a land of logging. The fact lives with us, supports many of us. We live with logging, by logging; many live for logging. It won't go away (even though it seems to be trying hard to do so at the moment). It is in the nature of the place, and to wish otherwise is as to wish the rain would go away: it won't happen here, not soon.

Logging was not the initial *raison d'être* of the arrivers. The immigrants came, mostly from Scandinavia or the Midwest, for cheap or free farmland and fish. To get to the soil they had to clear the massive forests that stood in the way. At first the big trees were seen chiefly as an obstacle to the rich agriculture everyone presumed would grow up in their wake. Then someone figured out that, since the work had to be done anyway, and since the farms often turned out to be stumpy or swampy, might it not be better to cut the forests systematically for timber and export it to San Francisco and other cities where it was needed for building? So this was done.

There ensued a frantic period of heroic logging, powered by mules, oxen, draft horses, steam donkeys, and men. Railroads entered and laced the hills with hundreds of miles of iron, ties, and trestles. Camps went up and the woods came down. The timber companies acquired vast holdings from the railroads, land grants, fake or abandoned homesteads, and an array of other deals. Boosters broadcast their pathetic hopes of the new cosmopolis as broad backs bent under the broadax, the misery whip, and the rain. And a lot of money was made, though not usually by the men who cut the trees.

This era, its wonders and spectacles, toil and spoils, feats and folly, has been amply documented in works such as Murray Morgan's *Last Wilderness* and Edwin van Syckle's *They Tried to Cut It All.* I've neither the knowledge nor the appetite to retell the true tall tales of that time. You've seen the pictures—twenty men in soiled long johns and drooping mustaches, or twenty-five women and girls in frilly dresses and hats, all perched in the slabcut of a huge Douglas-fir. You marvel at the spectacle and romance at the same time you weep for the giant trees.

Local heroes emerged. In fact, the only son of Willapa to be named to the Washington Centennial Hall of Honor was a logger: Oscar Wirkkala, a Finn who logged out of Deep River, was named to the elite list of one hundred luminaries for his development of equipment that led to aerial logging: a new way to take the trees. You might have known it would be a man from Wahkiakum who would be feted as a great logger.

There certainly *was* romance to the old logging days, and humor, heroic ambitions, and endeavors, as well as much pain, misery, and death. But there was also profit: large profits flowed from the bodies of the felled behemoths into the ledgers of the Weyerhaeusers, the Bloedels, the Simpsons, and their ilk. New behemoths grew in the place of the trees or, rather, in corporate buildings that scraped the sky where trees used to tickle the low clouds.

Meanwhile, the great old-growth timber proved exhaustible, a fact few were prepared to admit, believe, or even imagine. There were those, however, who looked into the cellulose ball and came away mindful of the future: a future where the fiber river ran dry when the timber bounty we'd taken for granted ran out like matches from a broken box.

As long ago as 1909, at a conference of governors in the White House, Frank H. Lamb of the Washington State Forest Commission wrote: "Western Washington has the heaviest and most uniform stand of timber in the world. It is primarily a forest-growing region; climate, soil and precipitation all are conducive to forest growth, but generally unfavorable to agriculture. . . . Originally we had about 300 billion board feet of standing timber in the State; 50 billion feet has been lumbered, about the same burned, leaving us about 200 billion feet remaining."

Lamb questioned even then the propriety of the process. "The homesteader who makes two spears of grass grow where one grew before is entitled to encouragement to the extent of free land," he wrote, "but under our land laws the lottery-like disposition of our resources has only bred a species of subsidy-seeking, graft-encouraging, perjury-promoting public spirit that has pervaded every department of our public life."

Lamb went on: "I hope that I have proven that the days [of our virgin forests] are numbered, that the hour glass is inverted. As surely as the grains of sand will seek the lower level, so certainly is the day coming when these forests, now the wonder and admiration of the world, the Nation's last reserve stock of timber, will be but a memory of the past; when the reverberating sound of the wielded axe and the roar of logging engines will cease to waken the once sylvan solitudes; when the smokestacks of a thousand mills, their days of usefulness past, their machinery gone to ruin, their thousands of busy laborers forced to other fields, will stand desolately forlorn, grim monuments of a past commercial era and a perpetual testimony to the heedless disregard for nature's treasures on the part of her servants."

The pioneers were not inclined to see themselves as nature's servants; quite the other way around. Yet the fact that Lamb would be proved right (as he largely has been) did not keep a few farsighted decision-makers from trying to head off the hourglass at the pass. Their results have been about as mixed as that metaphor. National forests and parks were created in an effort to save something back from the timber barons. John Muir and Gifford Pinchot clashed ideologically over given issues of conservation, yet their sparks led to national parks as well as national forests—places where the saws were to be stilled or slowed, respectively. Teddy Roosevelt created the basis for Olympic National Park and Franklin Delano Roosevelt expanded it, ensuring that something of the grand lowland rain forests of the maritime Northwest would be saved. If not for those moves, our forest heritage would be gone today. None of this occurred in Willapa. One of the greatest forests on the grizzled face of the earth, and no national forest or other reserve came to its rescue. A bit of state forest school trust land was designated here, the only appreciable public lands in the entire Washington Coast Range south of the Olympics: but under the dictate of state law, the Department of Natural Resources was charged with producing the maximum income from state-owned forestlands. This meant that the DNR administrators were forced to become timber barons in their own right, stripping state forests in such a way as to maximize profits if not long-term prospects on behalf of the people of Washington. Virtually all the rest of the forested uplands came into the ownership of Crown Zellerbach, Weyerhaeuser, Longview Fibre, and a very few other large companies.

As the timber giants were amalgamating their properties, an ideology of exploitation arose to justify their every action. As far back as 1923, apologists could be found for the supercutters. In that year, in *Basic Industries of the Pacific Northwest*, University of Washington business-school dean Howard T. Lewis wrote this odd piece of circular reasoning: "The Pacific Coast region needs a prosperous lumber industry and abundant forests. Without a sufficient lumber industry the new timber crops that can be raised on the non-agricultural and non-grazing lands will have no value, and without the raw materials produced by continuous wood crops we can have no lumber industry of importance to the future." Come again?

Lewis went on, "This does not mean that a curtailment in the production of lumber to husband the supply would be a good thing. On the contrary we should develop our lumber markets to the fullest extent so that the old virgin forests . . . may be replaced by vigorous young stands. If this old timber were gradually removed and the new growth established the land would again become increasingly productive. The problem can only be met by a proper encouragement of the use of wood and by immediate provision for reforestation. When it is known that the present stand of timber may be exhausted before a new crop of merchantable timber can be produced, the problem becomes particularly urgent."

At least Lewis recognized that the forests were exhaustible. Armed with convoluted rationalizations such as this, the axmen went to work. Yet reforestation did not match the cut; and the supply was certainly not husbanded. Nor could one really have expected anything different in the region, without any federal or state restraints whatever: after all, logs were life in Willapa.

And so the trees were cut. Of the old growth, next to nothing remained. And in many places, such as the Timbered Tor, a second or even third harvest has been taken. This has sustained, until recently, the towns and villages that arose from or adapted to servicing and housing the loggers and to moving and milling the wood they cut. As agriculture declined in the region, the people came to depend more and more on the economy generated by the logging industry.

But now the second growth is largely gone too, what's worth cutting. A recent study in forest trends shows that nearly all of the standing softwood timber in the Cathlamet district belongs to age classes between zero and eighty years. In fact, throughout the region, most of the trees are well under thirty years of age. Such sticks can go for chips but they don't make sawlogs. Nor can they support much of a local economy, especially since the wood-products industry is nearly moribund today.

So the wood-and-paper products corporations are pulling out, but when they've been the self-imposed only show in town for so long, it's difficult to back out unnoticed. As they go, the timber towns are hurting—really hurting. Why? If the trees came back for a second cut, why not a third? Along with other putative forest reforms came "sustained-yield logging" and "tree farming"—what has become of those admirable practices? Were they ever, indeed, put into practice, or were they just good ideas?

Here's where the story gets complicated. For in addition to the debate over how to manage forestland for long-term benefits and who's to blame for the timber turndown, there comes this inescapable fact: the land has been damaged along the way. Perhaps what follows should properly be accompanied by voluminous two-sided testimony, numerous citations, and statistics to back my assertions, as well as analyses of all the viewpoints. But this is no symposium. In fact, treatises on the subject exist, and they tend to back me up. Or one may consult several excellent books documenting the topic, among them *Clearcut: The Deforestation of America* by the propitiously named Nancy Wood (Sierra Club, 1971). Just as aptly named, syndicated columnist Steve Forrester has written penetrating columns on the ins and outs of the timber industry's malaise. Or read *The Forest Killers* by Jack Shepherd (Weybright and Tally, 1975). They'll give you the numbers. I am not so much concerned with statistics as with intense personal experience.

Nor do I pretend impartiality. In spite of two advanced degrees in forestry, I have my bias, revealed in the title of this chapter. All I can do is stick to it, ride it down, defend it, and follow it through. My perceptions and observations make up the meat of my story, and Mr. Weyerhaeuser can fire his figures in defensive volley if he cares to. All I know is what I read in the woods that have become newspaper. Or to put it another way, what I know is what I see—and I see the signs of the sack of the woods.

After the first cut, much of the land lay burned over and barren through the Depression. When a lot of it was sold for taxes, the companies were able to consolidate their holdings still further. Honest efforts were made to bring it back, on the part of foresters and nature. Erosion and soil compaction had not been so great with the oxen and the donkey sleds, not to mention the mileage of road building. Recovery happened fairly fast in many watersheds, at least in basic terms and on shallower slopes. The complexity of the primeval forest could not come back quickly, but in forty or fifty years trees stood tall again over many of the former stump fields. And, as planned, the loggers then moved in again.

Fair enough. Trees as a crop. We'll buy it. What went wrong? This time, no oxen, no railroads. Giant trucks. D-9 Caterpillar tractors. Loaders. Skidders. Bulldozers to scrape out the roads for the rest to ride on. An arsenal of ponderous toys with which to play pick-up-sticks in the forest. Sure, they were economical, and you can't stay progress on the logging side. But these new machines required tens of thousands of miles of new logging roads and spurs and platforms, and when they got there, they mashed down the already-poor lateritic soils into a hard, red glaze or exposed them in thousands of acres of crumbly cuts. In these roadcuts the loose, pebbly stuff rolled and eroded downward, leaving an arid, unstable slope unsuitable for most plants; the most adventive European weeds might grow here, along with horsetails, club mosses, the tough evergreen violets, and lower plants such as rocks invite: certainly not forests.

Meanwhile, the managed forests—the intensive forest—the high-yield forest—the forest people—the tree-growing people—the tree farms proceeded apace. What it meant was steep-slope, mechanized clearcut logging, followed by obnoxious slash-burning, followed by some replanting either by helicopter seeding, ground-level planting, or chance together with aerial spraying of herbicides that prevented the regrowth of nitrogen-fixing but competing alders and brush, fertilization to try to replace what's been lost, and (in theory) the creation of an even-aged, single-species stand of (preferably) Douglas-fir or (in Willapa) western hemlock: a boring, unstable, artificial monoculture.

All of this is justified by one or a thousand studies. Clear-cutting is deemed necessary to the reforestation of shade-intolerant Douglas-fir. Herbiciding is necessary, so they say, to create an economic density of trees (which must, nonetheless, be thinned later). The bears and ungulates must be heavily hunted to prevent their depredations on the young trees. It's all in the book. The many roads and spurs are required for economic access. And as I am told at county planning sessions when I raise the issue of erosion, "There isn't any erosion in these hills—the companies are doing a good job up there."

In fact, all I have to do is ask them—they'll tell me. In a publication entitled *The Environment: A Crown Zellerbach Commitment*, I read that "the company has developed a series of environmental standards for its woods operations to protect soil cover and watersheds and provide the conditions for the earliest possible regeneration of the forest." *Au contraire*—the way I see it, what Crown makes hereabouts is growth-resistant laterites, *ipso facto*. When I take visitors through Crown lands in Wahkiakum County, they sometimes ask me to stop—they feel the need to vomit. *The Environment: A Commitment*? I'm more inured to the sights, but that title makes me ill as well.

That particular publication is absolutely typical of the gaping gulf—the yawning universe—between nearly all timber corporation public-relations material and the reality I see on the ground. The apparent hypocrisy of the industry at corporate levels shows itself nowhere better than in the glossy fairy tales it gives out to schools, Granges, Rotary, anyone who will take it, to present its misleading view of the situation in the forest.

I, like most people, have been exposed to this misleading flummery for most of my life. Lately I've been collecting it, just for fun. In my file I find, in addition to the above-mentioned piece of Crown apocrypha, a beautifully designed Georgia Pacific booklet entitled *To Grow a Tree*. It tells me that "wildlife abounds" on their leavings, and that logging leaves "little if any erosion." The document carries a redwood branch-and-cone motif embossed in its Key-lime cover. It was produced shortly after GP logged redwoods around the clock in critical areas in an apparent attempt to sabotage the proposed Redwoods National Park. In the process, they felled the only tree ever known to top four hundred feet in height. The booklet ends with a litany of the tree species of America—just a straight list—as if not God, not Evolution, not Pan, but Georgia Pacific were responsible for their existence.

In a packet from the American Forest Institute I find a booklet entitled *It's a Tree Country*, which paints as pretty a picture as you could want of logging. "Man and Nature Work to Assure Tree Crops," reads the cant. Note the priority. The National Forest Products Association makes its entry with a brochure named *The Clearcutting Issue*. It cites all the tired reasons for relieving a stand of all its trees, dismissing the research-supported shelterbelt method as inconsequential. It informs us that clearcutting is nature's way; that the forest manager's greatest interest is to protect the soil; and that some clearcuts look "more 'virgin' than a virgin forest." I need not comment on that. How dumb, really, do they think we are?

Well, I guess we are dumb. In fact, in high school I was even dumber than when I wrote that letter to Crown. So dumb that I believed the Weyerhaeuser cartoons that have since become my favorites for their sheer gall. In those days I pinned the PR on my wall as inspiration for my planned career in forestry. We live and learn.

A pre-zip-code, pre-video Weyerhaeuser flyer from my file advertises the film *Tomorrow's Trees* as "a cinematic achievement" and "one of the most thrilling motion pictures imaginable." Subtle. (Intriguingly, the film is twice recommended as suitable for adult groups; perhaps "wildlife of the forest in its natural habitat" and "the towering majesty of nature at its best" are not what they seem!)

The accompanying pictures show a bear cub, watched over by its mother, the two of them surprisingly having survived the bear hunters Weyerhaeuser employed at the time, sniffing a precocious flower in the snow, snug within a tidy clearcut where the snag in which they'd hibernated was somehow spared. I can just hear the music and the narration of the film. The unsuspecting viewer would naturally elevate Mr. Weyerhaeuser to his pantheon of forest gods, placing him right up there with Robin Hood and Paul Bunyan. Somehow I doubt that the rest of the rustic deities would have him.

The surrealism of the old Weyerhaeuser style, I am forced to report, has not exactly been sacrificed to the whole truth in their modern offerings either. The 1983 company calendar shows all the sights we are likely to encounter in the "high-yield forest": goldfinches basking on a blooming rhododendron; prancing striped skunks, carefree in the sylvan setting of the clearcut; a young bear flipping a trout out of a stream in view of landscaped green clearcuts, beside a startled logger (he should be startled—knowing that company policy remains antibear to this day); a doe and fawns, peacefully drinking from a stream beneath a star much like that of Bethlehem (the month is December), which shines its approbation down upon the managed forest and its blessed dwellers; four quizzical otters, a family of juncoes, and a bunny rabbit all benignly regarding the beholder from a streamside in a mature forest (this one telling, perhaps ironically, more than it intends: for I can read in the otters' expressions, "What does that orange flagging all over the place mean?"); mountain goats on high surveying with obvious satisfaction the timbered land below; and a bald eagle feeding young on a nest above a landscape of little clear-cuts—never mind that the companies fight tooth and talon against practically every snag and eagle-nest buffer sought by the wildlife authorities on behalf of this endangered species.

It goes on: raccoons, fat from the fruits of the tree farm, play with tiger swallowtails, having nothing better to do, as ranks of perfect young firs grow up over lawnlike grass among abundant saved snags. (I suppose I should be grateful to see butterflies included in paradise, but tiger swallowtails, I curmudgeonly note, assiduously avoid both clearcuts and coniferous monocultures.) These pretty pictures bear no relation to reality.

My favorite Weyerhaeuser pinup shows a big mamma puma recumbent on a broad rock ledge. Two green-eyed, spotty baby cougars tussle over her broad paw, while a third peers coyly from her shadow. A hawk sails by and the mother mountain lion shares eye contact with the raptor.

Her expression can only be described as a beatific smile. She and the hawk must be sharing benevolent thoughts (we are to infer) about the charmed land in which they live, the joys of motherhood, the sweet life of the wilds (when slightly tamed), and the benevolence of their landlords. From the small, gentle-slope clearcuts in the background, we can tell that this is the "high-yield forest." From the contented-cow look on the pretty kitty, we can also tell that the tree farm is a great place for wildlife to live the good life. I wonder if the wildlife has been informed of this?

These classic ads make me want to create my own calendar for the company. I would use stark photographs from the grim depths of the Willapa Hills and captions such as this from David Wagoner's poem "The Lesson": ". . . a familiar forest/Clear-cut and left for dead/By sawtoothed Weyerhaeuser." But Wagoner's forest elegies I found on a remainder table in a city bookstore, while the Weyerhaeuser publications—which I see as fairytale forestry—go out in the millions of copies annually. That people prefer the comforting pap of slick PR to the disturbing poetry of reality would seem to be the lesson.

If Weyerhaeuser has remained as sophomoric as ever in its appeals to the public (probably finding such an approach effective), other companies have grown quite sophisticated. Boise Cascade (only modestly represented in Willapa) has been running ads recently showing the forester as wizard—conjuring trees and tree products out of nothing at all, to the astonished delight of children who, thereto, had wondered where such things came from. From Boise Cascade, obviously, and all the rest of the "tree people." Of course, the survival of the true pumas and real raptors does not depend on the company's commercial artists, but on its executives and shareholders. The fate of the forests themselves comes straight out of the corporate boardrooms. The directors being wizards, as we have seen, perhaps that's as it should be. The board giveth, the board taketh away. Only this time they're going to need more than a little legerdemain to produce trees on some of the damaged lands they've left behind.

One of the wizard ads invites parents to bring their kids out to the woods. "Show your kids how a forest is logged," it offers. "You'll discover how lumberjacks and foresters work with computer programmers and lab scientists to sustain our forests and their yields forever. Your kids will go 'Wow!'" I don't doubt they would: logging is pretty interesting. But I could take them to some other sites where their "wows" would have an entirely different sense behind them. To a Boise Cascade forest in the Cascades, for example, where the company decided to make a park for birders because of the diverse and abundant avifauna of a fire-damaged valley on Wenas Creek. I used to go regularly, with many other bird-watchers, for the spectacle. Then one spring we found that Boise Cascade had "cleaned up" the area, ostensibly to make it a nicer park. By removing many of the snags and dead, hollow trees, it also did away with the feeding and nesting sites of the woodpeckers, bluebirds, nuthatches, wrens, and other birds that depended on the deadwood and brought the birders. This wizardry I saw for myself, and when I did, I went "Wow."

If it seems arrogant for Boise Cascade to depict the blessings of the forest as strokes from its magic wand, the height of pretentiousness appears before Georgia Pacific's former headquarters in Portland. GP commissioned two sculptures to represent their corporate image. The first, known as *The Quest*, depicts three nudes afloat on the ether, in search of something higher and better. Known among Portland sculptors as *Three Groins in a Fountain*, they appear—and I am not surprised—to be looking away from the building. Claire Kelly, director of Portland State University's art and architecture program, described *The Quest* as "a very shabby piece . . . an expensive mistake." But as art critic Paul Sutinen says, "It is impossible to ignore," in which respect it does represent much of what the company has committed.

A poll to determine Portland's worst public sculptures conducted by Jeff Kuechle for the *Oregonian* and published in that newspaper yielded these comments. The critics also panned Georgia Pacific's second corporate work, which bears the modest title *Perpetuity*. Kuechle describes it thus: "a hollow cross section of a monstrous cedar, speared through with twisted splinters of bronze. In the center of the piece is a withered brown seedling. The colors are grim and drab, the theme vaguely disconcerting to anyone with an ounce of environmental sensitivity." Kuechle's colleague Alan Hayakawa pointed out that *Perpetuity* might not be entirely inappropriate, making "a great corporate symbol for a company in the logging business."

I'll say. Both works were commissioned from sculptor Count Alexander von Svoboda. Kuechle speculated that "von Svoboda, a Canadian, must have been a major GP shareholder to so blithely foist . . . his works on the giant wood products company." Perhaps it was partly to escape these embarrassments that Georgia Pacific retreated to its new headquarters in Atlanta, a move that foreshadowed a general exodus of corporate commitment (if not cutting) from the Northwest woods. At least they didn't bother to take the sculptures with them. They'd be better off going back to banal calendars, which would in any case be much cheaper.

Just as Boise Cascade maintains model sites where it delights in giving tours, Weyerhaeuser and the other biggies maintain demonstration forests along stretches of U.S. 101 and certain other heavily traveled routes. Here the lush regrowth admittedly resembles that on the calendars (minus the frolicking bear cubs and pleasant pumas). These picture-postcard forests boast broad buffers between the road and the rubble, smaller clearcuts than usual and on gentler slopes, intensive reforestation, and interpretive signs telling when a given stand was last harvested and replanted. They have been fertilized and cultivated to the state of the art, and they look pretty good. These exceptions to the norm are the picture a lot of travelers take home with them, making the calendars and ads seem all the more plausible.

But let me take you back into the hills, where tourists seldom go. There you will see miles and miles of scenes no company would dare advertise on a calendar or otherwise. Vast, interweaving clearcuts stretching ultimately for many thousands of acres; cuts that straddle stream and ridgetop, making no allowance for the limitations of the land; cuts that run steeply up one slope and down another, taking the slope out with the wood when the rivers run brown in winter.

Here is a passage from my journal, notes recorded during a crossing of the Willapa Hills that I made several autumns ago. "Such is the butchery of the clearcuts and accompanying erosion that, in places, waterfalls and cascades follow stone runnels unbordered by vegetation of any kind—like blood pouring over bones—bare, unbuffered runoff unable to be absorbed, to nourish and slake. Strange hills. The chopped knob of K. O. Peak has a knockout of a view of the Cascades and Olympics, but the near hills look shaved, butchered and lobotomized. . . . Down to the Willapa Valley groaned the most devastated hills I've seen, eroded brown mounds with half-burnt slash and virtually nothing growing, whereas other cuts on this side of the divide are grassy or weedy, and most on the Gray's River side are coming up in elder and vine maple or foxglove. Clearly, there is no average way for a mountain to respond to torture."

Many of the cuts are so ugly that first-time viewers sometimes retch or weep as they ask, "How can they *do* that?" They can do it because they've always done it however they wish, and they're not about to stop now. "But isn't there such a thing as a Forest Practice Act?" they ask. Yes, but it was written by a legislature dominated by big timber interests, and it has more loopholes than an afghan. Basically it says don't do this, this, or that, unless it costs too much, or unless merchantable timber is involved; in which case, never mind.

The word is that the standards have just been toughened; but you can't temper rubber, and besides, it's a little late. I'm a skeptic: I'll believe it when I see it. Substantially, there is little or nothing in the law to protect the streams, soils, and future forests from the worst effects of intensive logging.

A major conference convened recently in Seattle entitled "Streamside Management: Forestry and Fishery Interactions." Observers considered it more than mildly surprising that the sponsors of the event, the College of

Forest Resources and the College of Ocean and Fishery Sciences, both of the University of Washington, were at last getting together to discuss the impact of logging on streams and fisheries. An interdisciplinary approach has been conspicuously lacking in the past. If a joint initiative is not forthcoming, *Seattle Post-Intelligencer* columnist John De Yonge suggests, the governor and legislators should "take a look at the training programs, with an eye to reform." I suspect they'd better start looking.

After the first day of the meeting, a news summary vindicated my bad attitude. One of the key speakers for the timber side had maintained that no scientific evidence exists to show that streams or fish are hurt by logging. I slapped my forehead, in exasperation if not surprise. That guy should go to work for the tobacco industry if he wants a career change.

I await the published proceedings of the conference with interest, but without great expectations. Professional schools in both disciplines receive much of their funding from their respective industries, which are not likely to take kindly to costly reforms. While fisheries have much to gain and might be more open to cooperation, forestry has a lot to lose. Reform might mean, for example, that the industry would have to accept a new Forest Practice Act, one with teeth in it.

I myself couldn't quite believe that the existing law was so impotent until I saw it in action on K-M Mountain. K-M is a low shoulder of the Gray's River Divide that separates the Skamokawa Creek and Gray's River drainages. State Route 4, the main north shore of the Columbia River highway from I-5 to the ocean beaches, passes over K-M, carrying many thousands of travelers annually toward the shore or the interstate. K-M and the hills around and below it near Skamokawa carried a fine second-growth stand of timber until recently. Then they were barbered like a Marine recruit.

About the time K-M was due for logging, Crown Zellerbach adopted an incentive program to cut corporate costs by paying workers according to production shares instead of by hourly wage. It worked to a degree, for a while, until Crown decided to save even more money by contracting with its own workers as if they were gyppos (independent loggers), making them responsible for their own costs, and finally, to pull out altogether. These decisions and their aftermath have been reported in the *Wahkiakum County Eagle*. I don't know whether the incentive plan caused the men to log with greater fervor than usual, or if it was all in the company's plan; but whatever their reason, they logged K-M and the surrounding scenic lands as if trees would be extinct tomorrow. Indeed, to anyone driving over the pass today, it looks as if they might.

Right down to the road, right across streams, right over the next ridge, that cut goes. The first great horned owl I'd heard calling in the county occupied a beautiful little hill in a pretty valley at the base of K-M on the east. They all went with that supercut: the owl, the hill, the scenery. I have never seen a better or worse example of the utter failure of the Forest Practice Act to protect a sensitive area along a public right-of-way and a watershed from the harmful effects of logging than on K-M. No effort whatever was made to leave a scenic buffer strip as along 101. When I asked a company official about this, I was told that Crown Zellerbach could not accept the liability from potential windthrow of standing trees onto the highway. But to me it looks as though, in retreat, it just doesn't care. If it had to cut the whole of K-M or not at all, then it should have been not at all; and if that meant a public buy-off of the timber rights, or a swap of state land, then so be it.

Okay, so the view down to the Columbia has been opened up; but K-M is now one ugly drive, and it causes travelers that the county seeks to lure for a visit to step on it, to get out of Mordor in a hurry. I mean, it is *ugly*. Everyone I've asked considers what went on at K-M to be a travesty, and that includes loggers and their families and the families of corporate officials.

In fairness, I should add that Longview Fibre Company was involved in the K-M cut as well as Crown. In fact, out of this operation came a marvelously instructive symbol: an attractive, routed wooden sign placed by Longview many years ago, calling attention to the presence of the

Longfibre Tree Farm and admonishing passersby to "Keep Washington Green—Help Grow Trees." Someone forgot to take the sign down with the trees, so it still stands, vaunting its message to the disgusted passersby, backed by a shattered landscape reminiscent of newsreel clips of scorched-earth perimeters in Viet Nam. This ironic juxtaposition, the tree-farm sign in the midst of the logged-off mess, makes a suitable counterpoint to the calendar fibs.

I know that logging was never supposed to be pretty. Actually, it can be close to attractive, as I have seen it done in Mexico and Austria. But that kind of logging won't go here, where big trees need shifting over rough land. It is bound to cause something of a mess for a while. However, attractiveness isn't really the point. Health of the resource is the point. And when a forest straddles a major public scenic route, reasonable effort to protect the scenic resource becomes an issue. At K-M, the public interest in the forest resource has been miscarried in every way I can think of.

Now, spleen vented, fingers pointed, perhaps it is time for me to say that I am not antilogging. I am, after all, a product of two forestry schools, at the University of Washington and Yale. I have worked in the woods myself. I read (and write) books and burn firewood and use forest fiber in many ways. And in an indirect way, I am dependent on the forest economy. Repeat, not antilogging. The trees on K-M were ready to harvest. What I object to is the way it was done, with seemingly no regard for other values of the forest. And that is my objection to much of the history of logging in Willapa.

Likewise, my beef is not with the loggers themselves. They are seldom in a position to render decisions on the ground. They have a job to do, and skills, and families and bills, and they've gotta live. Nor is my quibble with the thoughtful managers in the companies, for there are some; I know a few. They believe in their mission, their profession; they may even believe their own PR. Rather, my gripe is with the decision-makers who should know better but still permit and promote the sack of the woods and with the stockholders who, by demanding maximum dividends, force untenable decisions in the office and unsuitable actions on the ground.

Of course, neither logger nor boss is likely to be sentient to some of the values I care about, and can't be blamed for that (after all, forest dwellers such as wood roaches and Johnson's hairstreak butterflies are fairly esoteric in their appeal). But a healthy, diverse, stable, and regenerating forest is in everyone's interest, biologist and logger alike. Unfortunately, it provides smaller short-term profits than short-cycle, high-slope cutting followed by monocultural management in some sites and no management whatever in other, forsaken places.

Make no mistake: they have done damage. Everyone says, "The woods come back." And to a point, they do. This land is the best for coniferous timber growth, the best there has ever been. The woods of Willapa were, in fact, one of the greatest forests on earth. The first cut, even with much waste, was the grandest anywhere in board-feet per acre, with the exception of the redwoods. The second harvest has not been bad. And—I admit it—in many places a fine new stand of fir or hemlock is coming along and will eventually furnish a part of our future fiber budget. Forestry after all, has its successes.

But it isn't coming back the same. There are reasons. First, the biomass and nutrients of an already rather poor soil base have been drawn upon so heavily that the fruitfulness of the hills has suffered. How could it be otherwise, between the removal of the cellulose skyscrapers and the burning of their branches? Empirically, the fertility must be down.

Nitrogen, in particular, has been depleted, most of it removed with the big trees, much of the rest rendered unavailable through slash burning, and replenishment prevented by herbiciding the alders. Nitrogen depletion may turn out to be forestry's downfall in these rain-leached hills.

Second, the remaining soil (such as it is) has been seriously eroded. I estimate that the Willapa Hills contain ten to fifteen thousand miles of log roads, spurs, and landings, or approximately fifty square miles of wasted, eroded, compacted, barren soil. The streams and their fisheries have suffered dramatically as a result, along with the woods.

And third, the complexity of the forest has been much reduced. The manager's ideal of a single-species, even-aged stand may make sense in the ledgers, but it leaves much to be desired on the land. Any monoculture under examination will be seen to be wanting the richness of parasites and predators that help regulate pest species and the soil fauna and decomposers that rebuild the fertility of the earth. This is especially true where a harsh chemical regime comes into play in maintaining the monoculture, as is true in the industrial forest. Lacking the usual issue of pests and their enemies, forests lose resistance, so that when any pest moves in it may thrive uncontested, and epidemic conditions may result. Then massive spray programs follow in response, costing dollars and further taxing the diversity, robustness, and stability of the forest. The managed forest of selected supertrees may produce more fiber in a shorter time under ideal conditions, but it stands vulnerable to decline and disaster in reality. This model is oversimplified, and it may not always happen this way; but it can. A diverse forest will be a healthier and stabler forest, and in the long run more profitable as well, than a monoculture.

I certainly do not oppose forest management for particular objectives. As Northwest Land Steward, I managed lands throughout the Pacific Northwest for The Nature Conservancy. Our "product" was natural diversity rather than wood, but I became highly sensitive to the need for management whenever natural processes and succession interfered with our objectives for the land in question. There can be no forestry without management of forestlands. But once again, I come down to *how* it is done, as well as where and how often. Smaller cuts, spaced over longer cycles, with minimal chemical manipulation and road-building, would mean better forests.

The logger may not care about the fact that a monoculture forest is a boring forest, from the naturalists' point of view. But he can relate to the ultimate threat to his job that overharvested, overmanipulated forests can bring about. Ironically, my plaint comes down finally to a duet with the loggers about their jobs, a blues riff if ever there was one this side of the Dust Bowl. For the logging industry is in the process of collapse in much of the Pacific Northwest, and no one in the corporate offices bothered to shout "timber!" or even "fore!"

The decline of the Northwest wood-products industry is a very complex story, and I do not pretend to understand all of its ramifications. It has to do with competition from Canada, whose mills are better tooled to deliver hemlock to Japanese specifications and are gaining a market edge from the strong U.S. dollar—you can buy Canadian two-by-fours cheaper than American ones in Portland, Oregon; a soft market for wood in general, leading to defaults on bids for public timber; a dramatic shift of logging action from the Northwest to the Southeast, where the companies have found they can make more money for now in long-needle pines with cheaper labor; and many other factors. One of them certainly has been overcutting. The accelerated cycles demanded by "intensive management" have left too much of the private land stumpage as just that—stumps. I hear it mostly from the loggers: "There isn't the timber up there; they've taken too much, too fast." The sack of the woods.

In other parts of Washington and Oregon, the pressure is on to increase the allowable cut on the national forests. A sympathetic administration responds by permitting money-losing operations on the public estate, and the liquidation of public old growth continues in order to shore up the sagging timber sector. One hears rancorous complaints about timber withdrawals for wilderness and national parks. These make dandy scapegoats for the malaise in the industry. An Oregon logging family patriarch carps, "Every time you turn around, there's another piece of forest closed off from logging."

Looking jolly in a hardhat in company with a Weyerhaeuser vice-president during a campaign stop, Ronald Reagan gave a pep talk to laid-off Oregon loggers as reported in the *Oregonian*. The president said that he could not support wilderness protection "in the wholesale amounts they are talking about. . . . Our private sector has not been guilty of rape of all the natural resources. There is today in the U.S. as much forest as there was when Washington was at Valley Forge." But what kind of forest? When he retires, he could do ads for the forest-products companies. We all know what Reagan said about redwoods.

Of course, when land is set aside, it removes the standing timber from the allowable cut. But the amount so reserved is minuscule compared to the acreage of land open to forestry, and in any case most of it exists at high altitudes where noncommercial species of trees dominate the scraggly subalpine forests. Little prime lowland forest has made it into the parks and wilderness areas. And when Mount St. Helens belatedly received protection as a National Volcanic Monument (it was the only Cascade volcano with no special status; in my view that is why it blew), conservationists succeeded in getting some unlogged forests included. In addition, the monument was to protect a large expanse of timber felled by the eruption, so that it might contribute to natural regeneration of the ecosystem. But before Congress could act, Weyerhaeuser made a killing from log salvage in the blast zone. Timber interests, blind to the need for protection of the greatest geologic phenomenon of our time, opposed the inclusion of any forest within the monument—standing or felled.

If all of the reserves were opened up tomorrow, it would not help an industry whose chief present problem is low prices for logs. Anyway, there are no such reserves in Willapa whatever, so the excuse just cannot be invoked here. No, the timber industry is in trouble, and it can't be blamed on the Sierra Club, the Canadians, the Arabs, or anyone else in particular. If there is any target for blame, it might as well be the companies, whose policies of liquidating the standing-timber base did not allow for a sustained flow of wood and wages. Blame, at least, for the manner in which the logging communities have been treated by the bosses—much of a oneness, it seems, with the way the woods were treated.

The companies have good reasons for cutting back: soft market, export competition, too much wood nationwide, too little wood in Willapa, some of it too big for scaled-down mills in Raymond, most of it too little for anything, ports and mills too far away and retreating farther annually, everything too, too unprofitable. And so the cutbacks and pullouts began, and with them the headlines: "Weyerhaeuser Shuts Down"; "End to Logging Means End to Paradise: Big Woes for S.W. Wash. County"; "Crown Cuts Forces"; "Crown Pulls Out." In the space of a couple of years, both of Wahkiakum County's biggest employers and landowners took a powder, leaving scores of families without a livelihood and several towns severely injured economically.

The closure of Weyerhaeuser's log camp at Deep River and Crown Zellerbach's operation at Cathlamet left an incredibly high percentage of the work force jobless—and with stunning suddenness. Used to high wages, the men had mortgages, kids, boats, RVs, TV satellite dishes, VCRs, and lots of bills. Now most of their houses are up for sale, with scarcely a chance to sell a one in the near future. The banks have begun to move in.

It was chilling to be witness to the slaughter. The talk in cafés was of drunkenness, depression, divorce, abandoned homes, Grapes of Wrath-type desperate moves in search of work, illness, expired benefits, crippled schools, damaged tax base, repossessions, and, occasionally, successful transfer or clever adaptation. There was also talk of the company; most often, the goddamned company. These families had been loyal to their employers, many for twenty years or more. Most families had suffered the death or severe injury of one or more of their menfolk in the dangerous occupation of logging. They couldn't understand why they were getting this in return. There was nothing to negotiate: the big boys held all the cards. They made paltry contributions to the Economic Development Commission, held conferences with the workers, and did what they intended to do. To myself and to other local residents, these looked like guilt payments and mock talks. Restitution was meager and compromise nonexistent. No one was surprised when the yellow trucks were gone, along with the jobs and taxes they represented.

The dust settles as after the felling of a forest giant, and the chips fall where they must. But these people's lives can't be dismissed by clichés, except perhaps in corporate boardrooms where mill and yard closures make clinical, bottomline sense. The communities survive, diminished, toughened; the people, mostly, survive, though many have to leave the land they love and where they chose to live. Life is preparation for change.

As one *Seattle Times* article had it, "Families will break up. Businesses will fold. People will move. The only natives likely to benefit are the elk and the eagle, shielded finally from the scream of the chain saw." All that came to pass, except for the silence, for the gyppos will continue to bid on state and company stumpage, insofar as they can turn a buck. The trees will grow where the soil remains intact, and in time, as markets change, the companies may be back. But timber will never boom here again.

Unnoticed among the sob-story features, perhaps, ran the following small headline: "Despite Poor Timber Markets, Crown Z Earnings Are Up." One of the reasons given by the then chairman of the board, William T. Creson, was "the improved organizational effectiveness of our people at every level."

Am I alone here in detecting a similarity to the logic employed in Crown's booklet, *The Environment: A Commitment*? "Environmental Commitment" is to the sack of the woods as "organizational effectiveness" is to the sack of the towns. And they both add up to cut and run. Life may be preparation for change for the common people, but for the timber barons, not much has changed at all. I wonder if it helps the out-of-work, out-of-benefits, out-of-hope loggers to know that the "improved organizational effectiveness" achieved by laying them off has helped keep the company's earnings up?

Faith in the timber companies had a lot to do with the failure to diversify the local economy when it might have been feasible to do so. Everyone worked in the woods, or depended on others who did. Then, as Bob Dylan wrote, "I had a job in the Great North Woods . . . then one day the axe just fell." Now, the ex-loggers wait for the trees to grow and the jobs to come back, as their unemployment runs dry in the rain.

It may seem to the reader facile and opportunistic of me to side with the loggers against the companies when my quarrel lies with the cutting of the trees in the manner in which the loggers have been employed to do it. But my quarrel has broadened. I live here, in a land of logging. My attitudes have changed since I came. I am more tolerant of the demands of forestry upon the landscape. I am able to edit (not ignore) the ugliness necessary to the logging economy, to an extent, in order to enjoy what's left. But I cannot edit the gross abuses that I don't believe are necessary. And I am a part of this community, albeit a relative newcomer. I am as abashed and offended by the callous crunching of these communities and their families, businesses, schools, and lifestyle as I am by the overzealous savaging of a wood. No, my quarrel is not with those who are here to make a living off the forest. I do the same, harvesting words instead of wood; and the pulp of the trees they cut, or others like them, mixed with their sweat, will carry these words. It would be churlish to oppose a professional activity that renders my own work possible.

I address my grievance, rather, toward those who dictated the destruction of almost all the original forest; who ordered the cutting of the Timbered Tor and places like it without the courtesy of discussion; who cut off the jobs of the men who logged the Timbered Tor, again without the possibility of appeal. In other words, toward all those corporate officials and stockholders (comfortable in their own feathered nests) who place profit above the people and the land from which they profit in the first place.

In February 1986, George Weyerhaeuser himself addressed mill and woods workers and their families in Longview, as reported by André Stepankowsky in the *Longview Daily News*. The president of the timber company that bears his name, the nation's largest, had come to ask his employees to accept cuts in their contracts of up to 45 percent in wages and benefits. Naturally, feelings ran high.

Refusing to stick to the subject, the exercised audience shot questions about the lack of company compassion for the widows of men killed in the woods; the absence of options for those laid off, many of them fired shortly before retirement was due them; how to cope with alcoholism and depression arising from being laid off; how to meet the family budget on lower pay than the men made a decade ago; lapse of a magnanimous company policy to treat twenty-five-year employees to coffee and cake or a free dinner; and other matters both weighty and pathetic.

But whether out of awe, or believing Weyerhaeuser's line that the company could not survive without the cuts, no one asked George the question I'd most like to hear him address: is he planning on taking a 45 percent cut himself? If he and a few of his directors and executives would do so, it ought to shore up the company's finances pretty well; and there would be no need to squeeze the workers any more. "In the last analysis," Weyerhaeuser said, "the customer is hiring all of us." Right you are, George. But somehow I doubt that he's worried about a pink slip in his paycheck or making the payments on his manse.

Ultimately, the workers struck. The company broke the strike, in the spirit of the times. And George got his way. The same week, Weyerhaeuser was named as one of the three worst polluters in Washington by the Department of Environmental Quality. It's nice to have good neighbors.

Decisions made a long time ago delivered the state of the present—decisions to liquidate the original timber base, fuel the boom towns for the present, and leave the future to fend for itself—along with the able midwifery of the midterm middle managers who similarly squandered the second growth. They say you should never spend your capital but live off the interest. That's what genuine sustained-yield forestry—real silviculture—is all about: living of the interest of the land. But when it comes to spending capital, there's never been anyone like the captains of corporate capitalism turned loose in the woods.

Looking long and hard at the literature on the subject, I find some cause for hope. Andrew Johns, of Cambridge University, writing in the *New Scientist* on wildlife's struggle to survive in heavily logged regions, concludes thus: "The timber industry and wildlife conservation are not mutually exclusive. The sooner that conservationists work alongside the timber companies to find ways of limiting the damage and preserving as much of the original fauna and flora as possible, the greater the prospects for survival of many rainforest animals."

Great, glad to hear it; but I think he's got the emphasis all wrong. As I think I've shown in this chapter and those that follow it, the above admonition might better read "The sooner the timber companies work alongside conservationists," not the other way around.

In a *Natural History* article, Richard H. Waring describes why the Pacific Northwest is the "Land of the Giant Conifers." Waring explains how conifers' shape, physiology, and ability to photosynthesize in winter contribute to their ability to thrive and grow immense in the maritime Northwest, rendering it wintergreen. The climate of the region and characteristics of the plants would seem to augur well for dramatic regrowth again and again.

But Waring sounds his own cautious refrain: "Perhaps the biggest threat to the giant conifers today is human activity." They dominate far less land than in Lewis and Clark's time, Waring says (Ronald Reagan's version of history notwithstanding!), and the processes that restore them—episodic fires, waves of hardwoods, and even epidemics of pests"—are excluded from both parks and reserves and commercial forests. "In human hands now lie the knowledge and responsibility to perpetuate—or doom—the largest forms of life this planet has ever known."

So it would seem that it may not be too late for intelligence and restraint to be applied in the woods. Wildlife and trees, according to these authors, have the resilience to respond to wise stewardship. The toughness of the people here tells me that they too have the stuff to survive, if given the chance.

But what of communities? Once disassembled, they come together again Humpty-Dumpty style: with difficulty. This goes for both natural communities (the elements of the ecosystem) and human communities. We'll not be here long enough to know about the former. As for the latter, the boom that came with the felling of the original forest led to expectations doomed to remain unsatisfied. The PR and promises of the timber companies promote similar false hopes today. The profit centers move, the woods struggle to their knees, the people evacuate or adapt. Tough trees, creatures, and people will survive. But the old woods, and the bright young towns of the log-train days, have been sacked.

What hurts most is the disingenuity. In the 1950s, Weyerhaeuser ran ads in Lower Columbia papers promising jobs for generations to come. In 1983, scarcely one generation later, a Weyerhaeuser vice-president spoke to the *Seattle Times* about the closure of the Gray's River logging camp. "It's almost more of a mental burden than I can handle to be the last big industry to pull out," he said. "There were once some pretty vigorous towns there. But the [Gray's River] operation is pulling the rest down." So it's Gray's River pulling Weyerhaeuser down, instead of the other way around. If the executive's sympathy is marvelous (I imagine he will handle the mental burden better than most of the loggers he left behind), his ability to shift the blame is virtuosic.

Ironically, the VP said that too much wood for too few customers necessitated the pullout. Others say the overcutting of quality timber, leaving acres and acres of young trees and stumps, caused the crash. Either way, Weyerhaeuser maintained, there was no more profit to be made. Yet when Sir James Goldsmith, the British financier, took over Crown Zellerbach's Northwest wood operations in 1985, his group had profit in mind. Former Crown woods in the Northwest and South, nearly two million acres of them, have been reorganized under Cavenham Forest Industries. In its press release of May 8, 1986, Cavenham president A. J. Dunlap said, "We intend to be a well-focused natural resource company and a competitive producer of quality wood products." How Crowncum-Cavenham/

Goldsmith actually plans to treat the woods remains something of a mystery. But based on the new act's first couple of numbers around here, it doesn't look good.

First, bailed out by Goldsmith, Crown copied Weyerhaeuser's vanishing act, orange trucks following yellow ones down the road. That gave eastern Wahkiakum County an unemployment picture to match that of Gray's River on the western side of K-M Mountain. Then the new company, in its first move as an absentee landlord, began contracting with gyppo loggers to provide pulpwood for its mills and peddling stumpage to other small outfits. One of the first stands to be cut was the Abe Creek Unit, a steep and beautiful drainage adjacent to a scenic stretch of the Columbia River highway. The huge clearcut that followed is now plainly visible from river or road.

Even a little company park, just outside the county seat of Cathlamet and for many years a favorite pleasure spot for locals and visitors alike, failed to escape the chain saw. Apparently no longer needful of local goodwill, the company included the park—name-labeled trees, rare yews, peaceful paths and picnic spots along with a small amount of merchantable timber—in the sale. A highway buffer one tree thick was left standing, the sheerest veil between passerby and travesty. Many were shocked. So was I, though I shouldn't have been; I've been there before.

Now I look across the valley at the scar that was the Timbered Tor and it seems less a reproach than a lesson. Foxglove has come to turn it purple in summer, vine maples redden it in fall. Eventually the surrounding woods will spread a graft of green over the torn slope. I will continue to miss the tall trees as the morning mist slips past, no longer pausing to dally in their crowns. As long as I am here, that cut will be a hole in the horizon. But it will become a more interesting hole as natural succession hurries to fill in the gap with whatever will grow.

Next door, the Grim Fell grows less grim each season. After the clearcut, after the herbicides, the new trees are coming in like a neat hemlock haircut. Having been spared the heavy erosion of many hills one can see from its summit, the old river terrace grows green once again.

The Grim Fell even shows signs of becoming good-looking in our time. Yet it lacks most of the interest and richness of Elk Mountain, across the creek to the east, where the alders, grand firs, spruce, hemlock, and other plants have come in helter-skelter, as nature saw fit. By contrast, the Grim Fell is a mere plantation; yet it seems to be a place where forestry is working better than in many others. It might even, one day, become a forest.

But how will these places be treated in the future? Will cutting cycles remain as brief as possible? Will the companies duck back in to make a quick buck as soon as today's young trees begin to mature, to raise hopes for a while before ducking out again and leaving the land and the people still more depleted than before?

Or will they begin to view the woods with a longer vision and the people of the woods as more than tools to pick up and put down and leave out in the rain when it suits them?

As forester Frank Lamb put it, concluding his 1909 address to the Governors' Conference: "Shall we see our children stripped of everything provided by a wise Providence for the sustenance of untold generations? The earth does not belong entirely to the present. Posterity has its claims."

# The Last of the Old Growth

Its tires complaining as they crunched over the basalt blocks that pass for roadbed in these hills, my Honda rounded the last curve on the logging spur. We came to a stop on a flat, bulldozed platform where a big orange boom had been positioned before. Way down below lay the short, narrow valley of Deep River: the straight, limpid stream, in high tide now; the roads on either side of it meeting at the white truss of the highway bridge and again a mile upstream at the shrunken remains of the once-booming town of Deep River. C. Arthur Appelo's original store and agency made a dull orange smudge in the middle of a small cluster of surviving buildings. Old hopes and old homes leaned into the deep green scene. Along the river, vital signs showed in the form of trim farms, sagging barns, and moored gillnet boats.

The object of the rugged drive up here was not the view of logging's legacy, the scrofulous hills all around and the semighost town below. It was something else the loggers left behind. As we got out of the car and strode a few steps across slash, there it stood: the Umbrella Tree—a relic of the old-growth forest, a single survivor, a lord without a liegedom. A grand (if lonely) tree, standing high on its clearcut ridge.

At Grange one night I was asking Carlton Appelo, C. Arthur's son, about old trees in the area. "You should really try to see the Umbrella

Tree," he advised. I said I'd never heard of it. "We always called it the Umbrella Tree, for its shape. All the growth is near the top. It's a remnant of the old-growth timber, a really huge Douglas-fir. They left it during the first cut as a witness tree, and I think it served as a spar tree too. The second growth has blocked the view, but I think it's still standing. You ought to try to get up and see it." Carlton, head of the telephone company and local ad hoc historian and archivist, went on to explain more about this survivor. A landmark until overtaken by fast-growing young firs and hemlocks, the Umbrella Tree had once again blended into the anonymity of the forest.

Carlton had heard that the stand was to be logged again, and he was concerned that the Umbrella Tree, as perhaps the oldest living thing in the county, should be saved. An old photograph, printed in one of Carlton's telephone books-cum-local histories, showed the tree on its denuded ridge following the first cut. It clearly demonstrated the origin of the name. Restricted to the upper part of the tree, the crown of foliage gave the appearance of a partially opened bumbershoot. I agreed with Carlton that the Umbrella Tree should be saved. But first, we needed to know whether it still stood: since the last cut, the second-growth forest had overtaken the ridge on which it stood, so that the Umbrella Tree could no longer be seen from below.

So one day in March 1984 I set out with Alan Richards, friend and local ornithologist, to seek the venerable plant. We left the Ocean Beach Highway at Svenson's Corner, headed up Oatfield Road and onto Rangila Hill, and followed the maze of logging roads in what we reckoned must be the right direction. Yet, although we encountered the only mountain lion either of us had ever seen, we failed to find the Umbrella Tree. Obscured among a great tangle of undifferentiated regrowth, it stood like some anonymous giant.

Since then the rumored logging took place. Carlton, County Commissioner Bob Torppa, and others worked behind the scenes to secure the protection of the tree once again. The owner, Martha Hess, daughter of original owner Bill Bakilla, agreed; and to their credit, Crown Zellerbach, who logged the stand, declared that anyone who damaged the tree would be heavily fined. No longer needed as a spar since steel booms came in, but still a good witness post for four sections and a spectacular example of the trees that went before, the Umbrella Tree was spared again.

So I decided to drive up there again. This time there was no cougar to be seen, but the Umbrella Tree stood out plainly against the pallid sky. When Thea and I looked at last upon its countenance, the Umbrella Tree stood alone again except for a few wispy saplings left standing around its base. In its solitude, one could truly appreciate the magnitude of such a tree: perhaps ten feet in diameter at its base, eight at breast height, and 150 feet tall. The crown had once been broken, perhaps by lightning, leading to a new double trunk that enhanced the unusual shape. I figured the tree must be very old indeed, perhaps five hundred years old, for all of its lower branches to have died and fallen from age and canopy competition for sunlight.

I gathered a cubic foot or so of orange-and-yellow plastic ribbon left to not rot all around the tree by the loggers, including one tattered slender banner looped around a slab of its scaly bark. On it in black marker read the command UMBRELLA TREE DO NOT CUT. The site thus tidied of garish flagging, I crouched at the base of the monolith to contemplate its size, age, history, and meaning. These attributes are hard to visualize, but the latter is the most difficult to make out. What is the meaning of a single old-growth tree: beyond being impressive and nostalgic, does it have biological meaning? Symbolic? And what of the forest that once, twice, stood with it and now is gone for the second time, leaving the Umbrella Tree alone in a field of stumps, overlooking a shaven slope? I was glad the Umbrella Tree remained. I felt affection as well as awe for it. But I was struck by the way in which the hoary survivor pointed up the absence of its peers—like some totem for the rest of the trees. At the Umbrella Tree, no matter how impressed you may be by its girth and age, you miss the rest of the trees. Happily, not far away, a bit of old growth survives where you might experience a whole forest such as the one to which the Umbrella Tree once belonged.

Searching for examples of the vanished original forest of southwest Washington, biologists of the Washington Natural Heritage Program came up with only one good candidate in the whole of the Willapa Hills. Known as Hendrickson Canyon, this 160-acre tract lies just twelve miles from my home and three from the Umbrella Tree. When we discovered the nearness of such a woodland we rushed to see it. But just as the Umbrella Tree at first eluded us, this patch of primeval forest proved difficult to find. We had maps and aerial photographs from the Department of Natural Resources (DNR), which owned the tract. On the aerials the old-growth remnant stood out like a pompadour in a company of chrome-domes, so bald were the surrounding sections. But the maze of logging roads that penetrated those shaven acres proved so convoluted and the maps so contradictory that the seemingly simple directions to the forest turned out to be most challenging.

In fact, the first several visits we made to the canyon turned out to have been only peripheral to the state-owned tract of old growth. Only when a number of local naturalists teamed with state biologists to comb the terrain did we manage to find our way into the depths of Hendrickson. It is easy, after such an exercise, to imagine how Sasquatch can live undetected in the tangle of successional forest around Mount St. Helens. It also shows how tiny the leavings really are, that they can be secreted among the immensity of the second growth, despite the size of the great old trees involved: just as the Umbrella Tree was swallowed up between cuts by the upstart youngsters around it.

When one frequently visits a special place, the recollections seem to run together in a psychic potpourri of encounters, sounds, sights, and smells. Seasons and companions help to sort out the walks and give personality to each pilgrimage. So when I think of Hendrickson Canyon, I think of which flowers were blooming, what birds were singing, and who saw and heard them with me. Augmenting my dimestore memory with the sounder bank of my field notes, I can separate the sensations into individual outings. It's like sifting out the hues from an abstract watercolor. A log of all our days in the old-growth patch would occupy another book, and I've got just a page or two for the purpose. So I will attempt, instead, a short set of field vignettes. Taken together, they may give some sense of the face of the place.

One of the earlier visits saw my wife, Thea, and me climbing an elk trail easterly into the old growth with a party of local naturalists. Botanists Cathy and Ed Maxwell and birders Alan Richards and Ann Musché were helping us begin to compose a biological picture of the place, preparatory to making a case for its preservation. Late autumn trending toward winter stilled the birds and muted the plants, except for a few. Bright chanterelles and even oranger polypore fungi daubed the secret places with unexpected color. Penny-whistling winter wrens shattered the silence that only shivered when the varied thrush gave its hoarse, monotonic call, which sounds like whistling through spit. Slanting great cedars hung draperies of moss mixed with their own lacy foliage, a green screen to the pinched winter light. Of that, little reached the mottled ground through the murky canopy of the conifers.

Winter was on the way out when I took my friend and mentor Charles Remington, of Yale's Peabody Museum, to see the old growth. We sought the wood roach, an enigmatic creature described elsewhere in this essay. We hoped to confirm its presence in Washington, where it hasn't been found for half a century. Broken, rotting bark never surrendered *Cryptocercus*, but it gave up a fascinating array of the invertebrates whose diversity gives the old-growth forest much of its richness. There were the many-legged myriapods, centipedes, and millipedes, those attenuated marvels of nervous organization. There were isopods, related to common "pill bugs" or wood lice, with their young collected around their legs in a display of parental care uncommon among invertebrates. And there were snail-eating ground beetles, their mandibles elongated to reach into coiled snail shells, and several species of their prey, notably the Vancouver green snail. Red-backed salamanders crept among the dark and rotting places in search of prey similar to our own. Charles collected a rich bag of insects indicative of the canyon's ancient and undisturbed systems. I promised to keep searching for the elusive wood roach.

Spring seems the liveliest time in the forest, and our expeditions center around that season. When our frequent field companion Fayette Krause came one March, we took to the canyon so that he might compare it to state park old growth he was helping to set aside elsewhere. Those turned out to be very different stands, distinguished by their immense Sitka spruces, the favored tree of maritime bluffs in Washington. While Hendrickson has big spruces, it is the truly mixed forest of the very old Willapa inland woodland—some spruce, some cedar, a lot of hemlock, and Douglas-fir. The largest tree we found that day was a massive Douglas-fir on the order of the Umbrella Tree's magnitude. So this is what its cohorts may have looked like, I thought, as we lounged by the big tree's base and pondered the wealth of the uncut woods: a treasure not measured in board feet, but in the number of notes in the winter wren's song, the legacy of leaflets in the carpet of oxalis shamrocks spread before our boots.

The next spring outing to Hendrickson found a band of old-growth pilgrims assembling at Swede Park on an April morning. Elizabeth Rodrick of the Washington Department of Game's Nongame Wildlife Program had come to see the forest as a possible habitat for spotted owls and other uncommon creatures. She brought her husband and son, Toby and Colin. The rest of the party included Carol Carver, Wahkiakum County extension agent; David and Elaine Myers, local photographer and master gardener; old-tree tracker Bob Richards, and others. We struck out for the canyon from a logging landing on the northeast and saw a whole new part of it. The hoped-for owls failed to respond, if they were there; perhaps we were too many in number. But we saw many another sight, including most of the local array of amphibians. A small, spring-fed stream took us into a spruce and cedar cleft where a pond stood, its water roiled by massing newts. Caddis fly larvae dragged their porta-cabin shelters across the gelatinous egg masses of the newts—feeding on what? So much remains to be learned in the forest, so little forest remains to teach it.

Our expedition reached a broad ridge completely carpeted with grass-green oxalis and overseen by big, burnt cedar snags from an ancient fire. Remains of three deer on this open forest freeway may have been cougar-kills; this was not far from where I'd once seen the big cat. Reluctantly we rounded the ridge and stepped back out to a clearcut, and cars.

Barely a month later, as May traded spring for summer, we returned with Fayette again. This time we found our way through the witch's maze of logging roads to the southwest corner of the quarter section that makes up the old-growth remnant. There is nothing natural about a square, and Hendrickson Canyon is surely influenced by land use (i.e., logging) around it. But fortunately, while far from a complete watershed, the DNR parcel contains portions of drainages on either side of a broad ridge, the shamrock runway we'd found before. This lends greater topographical and ecological variety to the site than it might otherwise have. During this May visit we explored that wide ridge northeasterly, gained the very point where the previous party had lunched and turned back, and peered down into the deepest, wildest crease of the canyon, where we have yet to venture.

Now the birds and flowers swapped their winter subtlety for spring's exuberance. Flycatchers, kinglets, and chickadees tittered as a western tanager drawled its tremolo call and a pileated woodpecker hooted and drummed. Chestnut-backed chickadees visited their nest fifteen feet up in an alder snag. Beneath, spring beauties, saxifrages, and wild lilies of the valley bloomed white against the jade hearts of the latter's deep-veined leaves. A different and rarer white flower bordered a beautifully fire-sculptured and lichen-painted cedar log. This pale blossom, one only to each stalk, has a different name in every flower book: single beauty, single delight, single-flowered wintergreen, single-flowered moneses, and wood nymph. The books describe it as lovely, charming, choice, delicate, and "one of the most attractive rainforest flowers," and they are right. But none of them mention its haunting fragrance. The procession of ivory, scented wood nymphs carried on for some distance beyond the log, flowering beneath young hemlocks that had arisen in the wake of the ancient cedar. An old tree falls, its crown gives up its sunshine to new young seedlings, a scarce wild-flower sprouts in the newly turned ground and feeds off the rotting wood. That's the nature of old growth: a chance to age, a chance to change, naturally and in the ripeness of time. Picking our way among an immensity of trees, we left by the way we'd come, with promises to explore that deep green cleft as soon as we could.

The year at Hendrickson came around with a late-autumn excursion in search of spiders. Our arachnologist friend, Rod Crawford, came to augment Professor Remington's invertebrate list with his own careful survey of spiders and other microfauna. In a brief circuit of one lobe of forest tangential to the main old-growth stand, Rod found twenty-two species of spiders, including one state record and one old-growth-indicator species; six kinds of harvestmen, a pseudoscorpion, an isopod, a centipede, and a millipede. This, after most of the natural activity had closed down for the season, and on a rainy day. We wonder what he might find on a fair day in high summer in the heart of Hendrickson Canyon?

Meanwhile, Thea and I found ourselves up to our ears in salmonberry and devil's club in a ravine we will likely avoid on future expeditions. For there will be many, I'm sure; we've never been in midwinter or midsummer, nor have we exhausted the roster of naturalists we hope to expose to our sylvan delight. As we do so, we build a better and better picture of a single old-growth forest, as such places grow rarer and rarer.

In Mount Rainier and Olympic national parks, I have hiked in old growth at places with names like Ohanepecosh, Hoh, Enchanted Valley, and Carbon River. Where the National Forest Wilderness Areas reach low enough to take in the deep woods, one can get their essence. These are about the only kinds of places where one can experience the old growth any longer—dedicated parks and wilderness areas, and a few *de facto* wildlands, almost all threatened, such as the Dark Divide between Mount St. Helens and Mount Adams. Almost everywhere else, the aboriginal forests have been removed and replaced with something else. So the survival of Hendrickson Canyon seems nothing less than miraculous in logged-off Willapa.

"So what?" ask wilderness opponents. "Trees are trees, a forest is a forest. When will you ecofreaks learn that trees are crops? Cut 'em, they'll come back; there'll always be enough trees to go around. Just don't lock 'em up!" Well, as we saw in the previous chapter, they may or may not come back. And even if they do, as they did on the Umbrella Tree's ridge, they won't be the same—not in our lifetime. Modern forestry has no intention of permitting the managed forests to gain anything like the stature of old growth before mowing them again. But let's be fair and answer these questions directly. What is so special about old growth? For that matter, what *is* old growth?

Our terminology suffers here from long confusion and imprecision. The concept of original, aboriginal, or primeval forest should be unequivocal, meaning as it was found by humans and undisturbed by them in any significant way. But there are almost none such to gauge by, even the Native Americans having had their impact; and natural fire, insect infestation, or wind damage often sets the clock back to the successional starting line without human help.

For the same and other reasons the term "virgin forest," a common usage, becomes unhelpful. For one thing, the sexual imagery involved in a metaphor of virginity necessarily equates all logging with rape. This intensifies tensions between those who would cut and those who would save, since foresters don't consider their profession to be ravishment. The metaphor of an unlogged forest as virgin is powerful, but as Gary Snyder suggests in this poem delivered at the Tilth Tenth Jamboree, the idea looked at another way is all wrong:

A virgin
Forest
Is ancient;
Many-
Breasted,
Stable;
At
Climax.

In Snyder's view climax forests are far from virgin. Besides, this metaphor for old growth inevitably leads to debates resembling late-night chat in a sorority house, concerning degrees of virginity and "technical virginity."

Virginity, like pregnancy or uniqueness, is an absolute quality. Old-growthness is not. Most forests have been disturbed to a degree, some quite a lot, then allowed to recover. Should they be excluded from consideration as preserves because they've had a burn or lost a few trees to insects, wind, or selective logging? At what point does disturbance militate against old-growth character? Or, putting it a different way, at what point does recovery mitigate disturbance?

As the debate over saving the last old growth heats up, several efforts are under way to craft workable and scientifically meaningful definitions. Symposia have been convened, cooperative research projects launched, and forestsful of paper offered up in documents aimed at clarifying the issue. In my opinion, some of the most valuable contributions have come from Dr. Jerry F. Franklin and his colleagues at the U.S. Forest Service's Pacific Northwest Forest and Range Experiment Station, Oregon State University, and cooperating institutions. Jerry Franklin is widely considered the guru of forest ecology and old-growth questions in the Northwest, and his research and ideas are widely admired. His pronouncements on the subject never pass lightly by.

In a weighty paper entitled "Ecological Characteristics of Old-Growth Douglas-fir Forests," Franklin and seven colleagues present the sum of their findings up to 1981. It would be a disservice to attempt to summarize these in the small space available here, and unnecessary. But some of their conclusions bear significantly upon what I have to say and should be repeated.

First, old-growth forest seldom exists in a climax state (Snyder's poem notwithstanding) since fire and the long life spans of the shade-intolerant trees involved (Douglas-fir) conspire to maintain subclimax forests for many hundreds of years.

Second, old-growth characteristics develop within approximately 175 to 250 years from catastrophic disturbance, depending on the nature of the site. Forests less than 150 years old should probably not be called "old growth."

Third, certain species of plants and animals occur only within old growth and would presumably become extinct without it; many more species find optimal habitats in old growth and may become quite uncommon if left to depend upon younger forests.

Fourth, most of the distinctive features of old growth depend on the presence of large, live, very old trees; large dead trees (snags); large down logs on land (nurse logs); and large logs in streams. These features create the basis for biological diversity as well as visual structure of the old woods.

So it is clear that old-growth forests do possess distinct features and values quite apart from maturing but still younger second-growth woods. Furthermore, Franklin et al. found that natural second growth (e.g., regrowth after fire or other natural catastrophe) differs dramatically from managed younger growth (i.e., industrial forests) in biological, structural, and visual terms. It should not be surprising then, when biologists claim special status for old growth. Since many of the organisms specialized for old growth are probably invertebrates and microbes, we lack adequate knowledge about them. But we can point to certain dramatic examples and say that without the old growth those would go.

There exists in old-growth woods in Oregon and the Great Smokies, for example, an odd, impressive insect known as *Cryptocercus punctulatus*, the wood roach—the one Charles Remington and I sought without success at Hendrickson Canyon. Possessing characters of both roaches and termites, this little-known animal seems to live nowhere else. Wellknown Oregon naturalist Stanley Jewett knows the insect in his state and feels strongly that it is virtually restricted to old growth. Populations formerly found in Washington State seem to be extinct, although we still hope to find it in one or another old-growth remnant. It may seem of little moment whether we should lose a creature with the traits of not one, but two, unloved groups of insects. But the intermediate, ambiguous nature of the animal renders it enormously important and interesting in evolutionary terms.

A larger, more lovable old-growth specialist, the northern spotted owl has drawn much attention in recent years for its picky habits. Rejecting other sorts of nesting grounds for sizable tracts of undisturbed timber, this reclusive and gentle owl depends for its very survival upon the maintenance of such forest. Thus it has become a precise indicator of fine remnant forest and something of a cause célèbre among wildlife and wilderness advocates.

By all rights the spotted owl should have been listed as endangered or at least threatened under the Endangered Species Act. But the resistance to such a rule-making would be so great from timber interests that the act itself might be jeopardized in Congress. Rather than risk such a confrontation, wildlife agencies and organizations have tried to work directly with private and public foresters to ensure adequate old-growth habitat for the owls. The debate has led, in the Pacific Northwest, to a spotted owl environmental impact study, currently under way throughout the Forest Service's Region Six.

Region Six headquarters had already directed each national forest to consider where timber withdrawals could be made to ensure survival of a minimal number of pairs of owls. SOMAs (spotted owl management areas) were to be three hundred acres in extent. In view of new research that suggests the birds may require more forest than that for their home range and for their young to find a home, conservationists challenged the

SOMA standard. That led to the current environmental impact statement, whose recommendations will surely be challenged by those who feel it offers too little for the owls as well as by timber companies believing the opposite. Because many logging firms and towns depend upon the purchase of federal old growth in the region to keep their crews and mills operating, the issue is a volatile one. A beautiful, inoffensive bird has become both rallying point and hated symbol for opposite camps in the woods.

The concept of managing the forests in order to include adequate owl habitat could be sound. Habitat suitable for spotted owls would contain viable populations of their prey, including red tree voles and northern flying squirrels. These rodents, while not absolutely restricted to old growth, find conditions there more suitable than elsewhere. The flying squirrel prefers to feed on truffles in the soil and fungi that grow on old snags, for example; this means that it and its predator, the spotted owl, do better where such snags occur: in the old-growth forest. Any reserve containing all these specialists would likely support much of the diversity indigenous to Northwest forests. Thus, by saving the major indicators, you can save much of the whole.

But if SOMAs should be too small, too few, and too widely separated, they would merely forestall extinctions by acting as island holdouts. The theory of island biogeography suggests that real or effective islands of habitat reach an equilibrium of diversity whereby new colonization balances routine extinction. Should there be no source of recolonization, islands will lose species over time, without replacement. This would be the likely effect of nominal, token somas established on a quota basis rather than on principles of scientific preserve design.

Spotted owl units have become virtually a currency in the old-growth controversy, being traded, sacrificed, gambled, inflated, deflated, and invested like so many stocks and bonds. But the security of the owl is far from insured as I write. And in the current administration, wildlife savers might as profitably invest in owl pellets as bank on the Forest Service and timber firms to exercise their option of saving prime owl habitat. They can't see the forest for the fees, and the owls lose out in the balance.

As a ranger-naturalist in Sequoia National Park, I led interpretive walks almost daily among intact stands of the largest trees on earth. After work, I would often return to walk or sit in the giant redwood groves, just to feel their presence and mass and to spy on the lives that went on among them. A pair of spotted owls nested not very far up in a burned crook of an ancient sequoia. I spent a long time simply watching the garganey-gray birds coming and going to feed their puffball young.

Nearby, in the very spot where John Muir liked to listen to the hermit thrushes broadcast what his friend John Burroughs called "the religious beatitude" of their song, I too tuned my thoughts to the panpipes of the thrushes. At such times as those, I was never sure just what was the best part of the old-growth redwood grove: the nearness of the plants and animals it supported, the grandeur of the trees themselves, or something about the very experience of the uncut forest that no parameters of wood or wildlife could define.

Certainly there is such an experience, to be gained in the ancient wood and nowhere else. If that were all we had to know—that old growth is something real, something different, that can be defined in scientific terms as well as by our muddier aesthetic definitions—then we could go forth, set aside all that's left, and be done with it. But while the above facts may be facts, we cannot so easily agree on the worth of the woods—the relative values of old growth as it stands or as it falls.

Everyone concerned knows that old growth furnishes a stunning, humbling, aesthetic experience. No one can deny that it supplies the only homes the spotted owls will buy. But it also provides some of the best high-grade lumber, at the highest densities per acre, and therefore the highest-priced stumpage to be found. And in most quarters, housing starts count for more than the clutch success of owls. So the trees continue to come down and Franklin predicts that at current harvest rates, the end of the unreserved old-growth forest is in sight within a few decades. The way that the annual cut has been accelerated under Ronald Reagan, it could come much sooner.

In fact, it has been the policy of timber-management entities to expedite that end. Their sole objective has been to recover board feet and bucks from the last of the grand old evergreens. The timber companies spent the bulk of their old growth long ago. We might have expected better from our public managers. Yet Bert Cole, long-time commissioner of public lands for the state of Washington, ran the Department of Natural Resources almost as if it were a subsidiary of that crosstown outfit that starts with "Weyer" and ends with "haeuser." In the process he implemented an explicit policy to liquidate old growth on state forestlands within this century. Serendipitous leavings such as Hendrickson Canyon hadn't a chance under that regime.

The United States Forest Service held back a little bit, depending on the political winds, but under the rule of Reagan that agency too sharpened its ax against the old growth and called for allowable cuts that no one seriously considered to be sustainable through regrowth. In fact, it was a Reagan Administration official who first made me understand the degree of contempt with which the timber-beasts regard noncommercial values of the forest.

The scene was the YMCA Camp of the Rockies, Estes Park, Colorado, during a convention of the National Audubon Society. Audubon's Charles Callison had taken on John Crowell, assistant secretary of agriculture in charge of national forests, on old-growth cutting policies. Crowell, fresh out of Louisiana Pacific's corporate vaults, addressed the usually mild, utterly respectable, and always moderate (not to mention largely Republican) Auduboners with remarkable truculence. He accused them of being radicals and zealots and allowed as how the spotted owl could damn well go extinct if it couldn't fit into the "proper management" of the national forests. Callison proceeded to demolish Crowell in his rebuttal and NAS president Russell Peterson drove home the stake. An embarrassed but unchastened Crowell slinked away to vociferous boos from the most worked-up Audubon audience I have ever seen.

Well, what do you expect when you have a timber-company executive in charge of the public forest trust? James Watt made ruder noises and more of them but may not, in his time, have done as much damage to the land as Crowell and his whole coterie of stoats in the chicken house. Yet Crowell's attitudes, sadly, are not uncommon among forest managers. While some recognize that "multiple use" means more than logging spelled five ways, many others see their proper role as extracting wood at the least cost and highest profit, and little more. And since old growth offers some of the highest profits of all timber, naturally they seek to cut it first.

Loggers operate under yet another compulsion to liquidate the old growth. The high profits to be had are one thing. But the fact that those big trees are out there just going to waste, dying and rotting, is quite another. Impelled by an anthropocentric view of resources from the start ("God put it there for us to *use*!"), they're driven wild. They can't stand to see all that fiber "locked up," unharvested, and subject to rot and disease that may render it unusable should they ever get at it.

Hence, the logging lobby has developed the idiotic concept of the "overmature" forest. By insisting that a forest must be logged as soon as it reaches maturity (in practice, often much earlier) or lose much of its utility and value, it hopes to scare policymakers into opening up more of the old growth to the saw.

With such arguments, timber interests have for decades tried to break into the rainforest corridors of Olympic National Park (see "Threads of the Green Cloth"). They have not succeeded there, but they have friends in high places in the Agriculture Department. The national forest wilderness inventory known as RARE II (for Roadless Area Review) demonstrated the federal attitude by excluding many deserving lowland forests from proposed wilderness areas or wilderness study status.

One can empathize to a degree with the loggers. With unemployment rising and mills closing down, frustration is high. Preserves look like a conspiracy to keep them out of work. Spotted owls and grandeur don't count for much compared to a paycheck. Conversely, those whose livelihoods originate outside the woods find it difficult to see why money matters more than beauty and ecological diversity. A lack of common values all around leads to combative positions and mutual insensitivity. A common bumper sticker in forest regions says it all: "Out of work? Eat an environmentalist!"

In reality, the woes of loggers do not originate wholly in old-growth and wilderness lockups. Most of the old growth has long been gone, most of the "lockups" lie at high altitude where small trees grow sparsely. And the high-grade low-land old growth is most definitely limited—it cannot continue to feed the loggers' families forever. They would be better served by proper management of regrowth on the cutover lands than by their employers' insistence on old-growth dependency.

Even the American Forestry Association has recognized the lack of wisdom inherent in the current policy. An editorial in the March 1985 issue of *American Forests* called for an immediate moratorium on old-growth harvest, to remain in effect until the ultimate allocation of this nonrenewable resource can be agreed upon. "If the only reason for harvesting is to keep afloat those timber industries and local economies that are on an inevitable crash-and-burn trajectory, then we say there is no reason to delay that inevitability with the sacrifice of never-to-be-replaced forests." This is not the Sierra Club talking—it is the major organization for professional foresters!

John De Yonge, a columnist for the *Seattle Post-Intelligencer*, points out that what is absent from old-growth politics "is sophisticated discussion about how to ease the economic shock that cutbacks in harvesting federal timber would cause." He continues: "It is not enough to say the national well-being dictates that certain entrepreneurs and their employees bear the costs of ensuring the United States maintains a full measure of all forest habitats. A society changing the game in the forests also should offer a pillow of transition for those who have to learn to play the new rules." Of course, he is right; and since the end of the old growth will come sooner or later, the proper course should be the development of a true sustained-yield forestry—something, as I suggested in the previous chapter, we do not now have. If we did, owls and others could have their old growth and loggers their jobs, and no one would have to eat an environmentalist.

So ultimately the argument should not be between loggers and conservationists, who both have a stake in the long-term productive management of the forest. Rather, the argument pits corporate officials and agency chiefs (often the same persons at one time or another) against people who apply a broader set of values to old growth. Such people know, or should know, that the "overmature" theory is specious.

In biological terms, there is no such thing as an overmature forest. Trees grow; trees die; logs rot; more trees grow. It is rather more complicated than that, but in essence the forest is a continually replacing, constantly refreshed entity. When the old-growth trees finally die, their crowns make room for newcomers to the sun and their bodies feed their seedlings.

In an ecological sense, a forest matures from about a century onward, having been young until then; but it never "overmatures"—Franklin's work shows that substantial net growth continues in the mature forest. Even if true climax were to be reached, a state of overall senescence would be prevented by fires, storms, insects, and fungi, always working to return portions of the mature forest to infancy and youth. The "over-maturity" argument considers only the value of the trees as they arrive at the mill.

Some cultures recognize implicitly that the forest has many more worths than the wood and have no such derogation as "overmature" in their forest lexicon. New Guinea tribesmen, for example, refer to the aboriginal jungle as "bikbush" in pidgin. It means big bush: the great rain forest, of big trees, big in every way. The bikbush provides bird-of-paradise plumes, birdwing butterflies, pandanus palm thatch and fruits, marsupial meat, spiritual succor, and many other benefits. It fits. Tribesmen who have traded the bikbush for timber royalties or oil-palm plantations miss these things. They no longer fit. We are not subsistence villagers; we make no headdresses from spotted-owl feathers. Yet we too would do well to recognize the many values of the old growth beyond simple timber revenues.

A companion piece to overmaturity is the myth that mature managed forests duplicate everything the old growth has to offer: a forest is a forest is a forest. Franklin's data, and anyone's eyes and common sense, show that this idea is balderdash. The spiritual awe to be gained from entry into a cathedral grove of giant trees, the homes for endemic animals and plants, the genes of trees selected by the centuries to grow biggest and best under given conditions, the inestimable complexity of a system that has worked perfectly since forests first covered the land—these and many other riches never come around again in the managed-forest cycle of one hundred years or less.

In dispelling the myth of overmaturity and related fables, we again discover the separate headsets worn by those who wish to keep the old growth intact and those who would prefer to cut it. One sings with the harmony of the complicated, interesting ecosystem, the beatitude of the big bush. The other jangles with the clang of cash and hums with the rhythm of rapid turnover of the forest resource. In the end, we just like the old growth as it is; they like it too—stacked up on an eighteen-wheeler, bound for the sawmill.

All this becomes almost academic in Willapa. The first half to three-quarters of the century saw the virtual removal of the original forest from these hills. Some of the last rumored bits on private land are said to be coming down now, as the big companies make their inglorious exit from the region. However, a movement is afoot to save a few fragments. It's our very last chance.

The most important piece in the region, indeed one of the most crucial unprotected old-growth groves anywhere, stands on Long Island in Willapa Bay. Though beyond my tidewater boundary, Long Island nonetheless belongs with the Willapa Hills. An outlier, attached at a time of lower sea level, the isle owes its ecological allegiance to the hills rather than to the maritime Long Beach Peninsula across the bay. The largest estuarine island on the Pacific Coast (six miles long, about 4,700 acres), Long Island suffered little change from the centuries of periodic occupation by Chinook Indians prior to the whites' arrival. A shantytown of oyster-pickers sparked briefly at Diamond Point on its north end, between 1867 and 1878, fueled by the nearby beds.

Timber extraction began after the oysters gave out, and the Weyerhaeuser Company acquired 1,600 acres of the island. The federal government added Long Island to the Willapa National Wildlife Refuge in 1939 and proceeded to acquire much of the remaining acreage. Somehow, a stand of ancient and giant western red cedars remained standing, sheltered within the deeper woods of the island's southern reaches. Unfortunately, the cedar grove lay in the timber company's inholding rather than on refuge land.

Aiming to consolidate its holdings, the Fish and Wildlife Service initiated a complex series of land-for-timber exchanges in the 1950s. These have continued to the present. In the most recent transaction, Weyerhaeuser received twenty-one million board feet of federal timber in exchange for 1,622 acres of its land. This was supposed to safeguard the giant cedars. However, when cedar prices climbed, Weyerhaeuser insisted on a reappraisal. This favored the company, so that the exchange fell short of securing the complete cedar grove. Eventually, 119 acres of the big cedars came to USFWS. The agreement contained a buy-back option on the remaining 155 acres, valued at $5.5 million. Weyerhaeuser's cedars were to face logging over a six-year staged operation beginning in 1987 unless funds for their purchase could be allocated.

Cutting any of the old-growth cedars would open the rest of the stand to serious windthrow danger and irreparably disrupt an utterly unique ecosystem that has remained intact for aeons. Yet, with Weyerhaeuser holding the cedars hostage, as it were, for the government ransom, conservationists nearly despaired at the prospect of procuring the first payment under the current economy. But the Wilderness Society, the National Audubon Society, and others named Long Island's cedars a priority, and the Willapa Hills Audubon Society tackled the challenge whole-hog. Irene and Steve Bachhuber, leading the effort, sought to gain support from residents in logging-dominated southwest Washington. Many of us responded with letters to congressmen and senators. Then a series of events strengthened our case.

First, volunteer Jim Atkinson located a pair of spotted owls in the cedar grove. Their presence furnished strong new evidence for the advisability of saving all the cedars—the stand barely meets the extent considered minimal for spotted owl breeding as it is, and any logging in the grove would almost certainly make it unsuitable for the rare birds and any associated fauna.

Next, a team of prominent conservation biologists visited the island on November 6, 1984. Led by Jim Hidy, refuge manager, who knows the island intimately, the delegation included usfws official Susan Saul, cited in private life as the moving force behind Mount St. Helens protection. The leading journalist on Mount St. Helens matters, André Stepankowsky of the *Longview Daily News*, covered the outing for the press. Among the noted botanists and ecologists present was Jerry Franklin himself.

Visibly and profoundly impressed, Franklin pronounced the cedar stand distinct from any other he knew of, and the last opportunity to protect what was once a widespread forest type on the Northwest coast. Quoted by Stepankowsky in *Washington* magazine, Franklin confided, "I really thought I was going to see something quite ordinary. I was surprised. You're so often disappointed. But in this case, I saw something extraordinary, something I've never seen before."

The other scientists agreed. The extent and uniform nature of the stand, they decreed, was unparalleled. Individual trees exceed ten feet in diameter at breast height and one thousand years in age, and the stand as a whole has not experienced a catastrophic fire or storm for the past four thousand years. Besides being ecologically distinct from any other known stand, the Long Island cedars comprise a rare example of a self-reproducing climax forest in the strict sense. Everyone had known the Long Island cedars were something special, but this visit showed just *how* special, as it stamped the imprimatur of solid science on the protection effort.

Finally, Congressman Don Bonker toured the cedar grove late in 1985. Representative Bonker had shown his environmental colors by steering the Mount St. Helens National Volcanic Monument and the Protection Island National Wildlife Refuge through the last Congress, the only major land bills to get past Reagan and Watt. Congressman Bonker too was impressed and promised to help.

He did. The fiscal year 1986 appropriations bill reported out of the House Interior Appropriations Subcommittee contained $3.4 million for acquisition of a substantial portion of the grove. But in the budget-crunched debate, the Senate failed to match the House's offering. Senator James McClure of Idaho, chairman of the Energy and Natural Resources Committee and an environmental nemesis, failed to include the appropriation in the Senate bill in spite of his Northwest colleagues' requests. Conservationists urgently requested Washington's Senators Slade Gorton and Dan Evans and Senator Mark Hatfield of Oregon to press for Senate funding prior to adoption of the final budget. A letter I received from Senator Evans, written November 6, 1985, reads in part: "There is increasing national and local interest in the old-growth cedar stand on Long Island in Willapa Bay. . . . I strongly support the acquisition, and will encourage adoption of the House provision during the House-Senate Interior appropriations bill conference." Senator Gorton concurred. He wrote that, despite his concern over the federal deficit, he believed that "this option must be executed, thereby saving this unique old-growth cedar stand on our coast. . . . I am committed to doing everything I can to support funding for the Willapa National Wildlife Refuge."

It worked. On December 19, 1985, Congress passed and President Reagan signed a bill including a $3.38 million appropriation to purchase seventy-eight acres of the remaining cedar stand. This was a great success and a testament to the educability of Congress and the power of popular opinion on behalf of nature. However, half of the stand remains in private claws and may be logged beginning in 1991 if it is not bought by the government by then. Congressman Bonker has managed a bill through the House to furnish funds for the other half, although he objects to the idea of paying Weyerhaeuser another inflated increment while cedar prices are in fact dropping. Getting another such appropriation through the Senate will prove challenging. Meanwhile, the Gramm-Rudman-Hollings Deficit Reduction Law automatically pared $150,000 from the Long Island allocation, leaving about two acres of cedars vulnerable to cutting as early as 1989. So the drama—and the fight—goes on.

Meanwhile, on the mainland, a quieter effort goes on to safeguard another old-growth crumb of about the same size. Searching for a remnant of the former forest of Willapa, the biologists of Washington's Natural Heritage Program ferreted out 160-acre Hendrickson Canyon. We visited it a few pages back. It is truly remarkable that such a forest should have survived in the logged-off land, more so in view of its location—just a mile from one of the biggest logging camps ever to work the Willapas. Carlton Appelo's *Deep River Story*, published as a historical supplement to the 1978 directory of the Western Wahkiakum County Telephone Company, gives some idea of the enormous amount of timber that passed out of the surrounding hills to tidewater at the Deep River landing. Yet somehow, this one-quarter-section of timber survived almost untouched.

Surrounded by private timberlands, Hendrickson Canyon itself belongs to the people of Washington. The Department of Natural Resources administers the state forests on behalf of a number of trusts. Hendrickson belongs to the university trust, designated to produce income for the five state universities. Apparently its stumpage has never been put up for bid, and so it stands today.

Some locals, including old-timers who have logged in the area for decades, maintain that Hendrickson is not old growth, having been logged by mule team around the turn of the century. But the Heritage biologists could find evidence of only four cedars having been removed, and we find virtually no stumps in the interior. Otherwise, they found no evidence of human disturbance. While a few trees may have been high-graded, by all accepted standards Hendrickson Canyon retains impressive old-growth characteristics.

The Heritage Program, a branch of the managing agency, recommended retention of the site by DNR as a natural area. Its report read, in part: "Although similar plant community types occur on Federal Research Natural Areas (RNAs) on the Olympic Peninsula, the assemblage of community types here is different from those of the RNAs. Furthermore, this site represents one of the very few remaining old-growth forest stands in southwestern Washington, and the soils, geological substrate, and landforms here are different from those of the RNAs to the north. These differences are sufficient to warrant preservation of the site."

Delighted by the prospect of an old-growth preserve so close to home, I went to work with others to seek support for Hendrickson's long-term protection. The regional Resource Conservation and Development District, the Wahkiakum County Democratic Convention, and finally the Board of County Commissioners all passed resolutions urging DNR to reserve the canyon as an old-growth natural area. Commissioner Bob Torppa added the thoughtful suggestion that the university's College of Forest Resources could utilize the site as a research and demonstration forest, thus providing the university with direct and valuable benefits without logging the trust land. (Other college forests lack significant old-growth stands.) The commissioners went on record as "supporting the preservation of the Hendrickson Canyon Old-Growth Forest to protect the same from being destroyed or severely impacted by the encroachment of man for uses other than educational and conservation purposes."

Some thought it strange that county officials in a logging-dependent district would unanimously pass such a resolution. Perhaps it would have been different had the land belonged in a local school trust instead of the universities: the county's share of revenues from logging on common school trust lands is way down due to the ailing timber economy. As it is, the county stands to gain little and lose much should the site be logged. They saw this. Even the loggers I've discussed it with support such a reserve as a heritage site where their children and grandchildren can see what the trees that built the community were like. Where else can they experience such woods anywhere near?

The fact that old-growth indicator spiders occur in the canyon, as my arachnologist friend, Rod Crawford, found, may not impress anyone in the tavern; but the local heritage idea makes sense. Dennis Nagasawa, neighbor, fire chief, and local DNR man, told me that cones have been gathered from big trees in Hendrickson for seed recovery. The genetics of that makes sense, too. Developed over the ages, the genes in those seeds are postcards from the Pleistocene, phone calls to the future of forestry: you can't buy them from Burpee's.

We have solid local support to save our last, best Willapa forest. That would have availed us little with Bert Cole in the cockpit at DNR, one hand on the controls, the other on a chain saw. But, happily, attitudes are changing at more levels than local ones. When Bert Cole tried to get away with a single Environmental Impact Statement for all the state's timberlands, and that pretty sketchy, the courts threw it out and the voters threw him out of office.

The new commissioner of public lands, Brian Boyle, came in with a fresh set of policies that place greater emphasis on protecting the unique qualities of the state forestlands. Among these, a policy has been adopted that will set aside a number of so-called "Old-Growth Seral Stage Deferrals." The stated rationale reads conservatively: "Seral Stage deferrals are for the purpose of retaining the option of acquiring information on old growth ecological relationships; applicable to intensive timber management." (Not for their own sake; but not to quibble.) Hendrickson should unquestionably be the deferral for DNR Seed Zone 041, and the rationale is less important than the fact of protection. The deferral, if granted, is supposed to last until 1993, by which time a way may be found to dedicate this special place as a State Natural Forest Area.

I do not want to imply that opposition to old-growth preservation is absent from Willapa. In sympathy with their profession, which is hurting all over, logging folk tend to be suspicious of reserves. But the fact that the area's old growth is on the whole long gone means that we lack too the virulent battles of the national forest communities, where the bone of old growth is still in contention. One of the few blessings, that, of living in a logged-off land.

Crown Zellerbach has opposed the Hendrickson designation sub rosa, apparently just on principal, or perhaps because of what it sees as the inconvenience of having a preserve plunked down next to its managed land. But who would have expected anything else? For the neighbors of Hendrickson Canyon to get behind the community in a common effort to preserve a minute example of the forest that built them both would have been too much like cooperation and entirely out of character. They have their image to maintain.

Similarly, Weyerhaeuser certainly hasn't made it easy to save the cedars of Long Island. Any measure of public interest whatever would dictate the donation of those trees for the tax advantage and public goodwill it would bring, and for the good of the people of the state and nation that permitted the Weyerhaeuser empire to flourish. But no; the corporation must have top dollar, when the nation can ill afford it, or it'll cut. When approached on the subject, Weyerhaeuser spokesmen have retorted that the company plans to survive and that they have more interest in doing so than in public relations. Yet, Weyerhaeuser owns some fifteen million acres overall, about 1,724,000 in Washington alone. The idea that the remaining eighty acres on Long Island will make or break the biggest timber giant of them all is definitely hard to swallow.

In his *Washington* magazine story on Long Island, André Stepankowsky quoted company spokesman Mike Munson as saying, "We feel like we've already gone the extra mile. We've forgone that asset on Long Island for a long time. We understand the value of the cedar to others, but it's of significant value to Weyerhaeuser."

I'll say. Weyerhaeuser made sure that the trade agreement contained a clause increasing by 12 percent annually the amount the government must pay for the grove—even though cedar prices are now falling. It's gone the extra mile, all right, but only in the race to the bank. As far as the public interest goes it hasn't budged an inch. Nor is it enlightened self-interest to act as Weyerhaeuser has in this case and others; rather, I'd call it benighted self-interest—unencumbered by concern for the land, the people, or the earth.

There are those who believe that the corporations haven't the right to doom the last of the old growth. Members of Earth First! and other monkey-wrenchers act as environmental guerrillas for the cause of "deep ecology"—the principle that the health of the land must come before corporate profits. They have been known to sabotage logging equipment, drive metal spikes into trees to make them impossible to mill, chain themselves to trees, and otherwise impede the loggers of old growth. Judging by their actions elsewhere in forests of lesser importance, one would not be surprised to see such people seek to protect the Long Island cedars by such tactics if all else fails. I personally doubt that those trees will come down without a battle. That the log trucks must get to the island by ferry-barge makes the operation particularly vulnerable. If it comes to that, it won't be pretty.

Personally, I prefer to arrive at conservation goals through civil means. Working with the commissioners, the DNR, and local citizens to save Hendrickson will have been a satisfying process if we succeed.

Likewise, Fayette Krause, Washington Land Steward of The Nature Conservancy, has worked closely with Washington State Parks officials to ensure protective management of old-growth trees and windthrown deadwood in a number of parks—including gigantic Sitka spruces at Forts Columbia and Canby state parks on the southwestern hems of Willapa. In the past, salvage operations took more deadwood and standing trees than was necessary for public safety and park operations. Now, acting on the advice of a citizen's committee, State Parks Director Jan Tveten intends to conserve old-growth values in the parks whenever possible. Several Natural Forest Areas have now been established in the region's parks.

In another example, state Nongame Wildlife Program biologists seek cooperation of landowners to protect buffer zones around bald eagle nests, such as the old-growth scrap at Brookfield Point on the Columbia River near here. Crown fought for every last tree, but the eagles remain for now on the steep, never-cut bluffs above the river. In this case, the Endangered Species Act was invoked to protect an old-growth parcel through mandated consultation.

The land trade on Long Island comprised a good-faith effort on behalf of the Fish and Wildlife Service and conservationists to cooperate with Weyerhaeuser. Even without the cedar, the company has made out very well on the swap, getting abundant marketable federal timber in exchange for their logged-over lands. After years of tedious attempts at compromise, corporate recalcitrance over the priceless cedars means frustration for many. The company's attitude seems designed to provoke desperate measures.

Still, it is better to get what old growth we can through peaceful pressure, negotiation, and use of the legal tools we possess than to try to force the outcome with violence, even though the effort be great. But when the timber giants hold all the cards, the options are few. Let's hope that Weyerhaeuser will see its way clear to be reasonable in its demands, or even to donate the rest of the Long Island cedars; otherwise the results could be expensive and tragic for all concerned. By upping the ante on a national treasure, some say Weyerhaeuser has asked for whatever it may get, be it dollars or dynamite. Let us seek our old-growth goals with peaceful guile instead and hope that the opposition will see fit to cooperate more fully than in the past.

I went to Long Island in the rainy spring of 1986 to view the cedars with a group of conservationists and to begin an inventory of the grove's smaller-scale wildlife. Our party roughly circled the set-aside core of the old growth, led through the shaggy vegetation by USFWS forestry technician Jody Atkinson. The way was marred by orange paint and flagging on the trees. But for once, the garish blazes indicated trees destined to stand instead of fall; and they will be effaced in the future. Still, as we peered through the foliage into the forest beyond the pale of protection, we could make out the bulky forms of giant cedars still slated for logging unless their ransom comes through.

We walked among cedars six, eight, ten, twelve feet through, their boles crumpled like metamorphic features of the earth. Brown hemlock bark and the gray scales of huge spruces contrasted with the red swirls of the cedars. I shivered at the thought of saws ripping into their ancient hide. Certainly, it was clear to us all that the grove as a whole would suffer if its frontier were to be violated. What's left is not enough, but it's all we've got, and nothing less than all of it will do.

The trail, such as it was, ran between a pair of enormous red cedars. These twins shared a base as well as the scars of an ancient fire that left their boles hollowed out. I stood between them as the group carried on and could have disappeared into either. These cedar caves brought back a pang of my boyhood desire to live in such a place—they had that sort of capaciousness about them. I could see that many creatures, if not boys or hermits, took advantage: their holes, burrows and borings, scrapes, pellets and feathers, all told of it. Each cedar is a city in itself; the grove, a rich, living nation of life that should not be divided.

Heading home that night we rounded Rangila Hill, and I thought of that other great high rise, the Umbrella Tree. Unlike the cedars of Long Island, the great Douglas-fir had lost its neighborhood. Standing on its windy ridge, the Umbrella Tree was the last of the big old trees to be seen for miles around. And as such, it could not stand forever. We'd thought we had it saved. Then in a great storm off the sea in January 1986, the Umbrella Tree came down in the night. Broken off at the base, its immense bulk fell far down the cleared slope below and broke up like a ship on the storm-beaten shore. The small victory achieved in the saving of the Umbrella Tree proved, after all, to be pyrrhic.

When the loggers took its surrounding shield of trees, the Umbrella Tree was left to weather the full force of the Pacific gale, unguarded. Windthrow proved its undoing, just as it would damage the standing cedars on Long Island should the buffer of their weather-proven edges be removed. But the Umbrella Tree seemed to be dying anyway. Its top turned brown after the logging around it. A poet might say that it mourned the loss of its forest for the second time in a century, and the loss proved too much to take; a forest pathologist, that the logging around it damaged its roots. Or maybe it was just that old tree's time to die. For even the old growth is not immortal.

Over by Skamokawa, Bob Richards is trying to save another famous solitary giant. The tall Douglas-fir may be seen from many vantages in the hills and valleys above the Columbia. As a fresh logging operation approached the ridge where the Landmark Tree stands, Richards asked for it to be spared. "To those with the power of life and death over this non-renewable resource," he wrote in the *Eagle*, "I humbly beg, please spare the East Valley Landmark Tree." In the hands of the loggers lie the growth and the untold tales and the evidence of the ages.

Biologically, such trees as these mean little except to the animals that live in their fissured bark and their boughs. (Among the plastic flagging scattered about the base of the Umbrella Tree, we found the bony pellets of great horned owls; how many generations roosted in that hoary crown?) Historically, they say somewhat more. And symbolically, they mean quite a lot.

Woodsmen call the lone, left-behind trees like these "wolf trees." As they grow rarer, that image becomes increasingly suitable. The big old wolf trees, while they still stand, stand for something. They can represent the senses we could still come to, too little, too late, but at last. Or they can be seen as lonely monuments to our stupidity, cupidity, and greed. It all depends on how you look at them. And on how we treat the last of the old growth.

# There Ought to Be Bears

Red Almer, versatile plumber-cum-publican and bandsman, erstwhile county commissioner and high-top logger, sat opposite me in the living room of Swede Park. A water-heater crisis averted, we settled down to an ale before the cookstove as an October gale ripped leaves off the oaks outside and pasted them to the windows like schoolroom decorations. Red's scarlet jacket and red Oasis tavern cap echoed the oak leaves and seemed to stain his paler russet hair beneath the cap. We talked about the early deer season. I asked Red whether he'd seen any bears this year.

"My hunting partner saw one last week," Red replied. "But that's the first one I've heard of for a while. Twenty-five years ago, I'd see a bear every weekend. We'd run into them in the brush, or hear 'em crunching apples in the old orchard up East Valley. But anymore you hardly ever see a bear. You can't find one if you want to." Red frowned at the thought.

Joe Florek, a current commissioner and a logger, told me a similar tale in a café conversation. "Used to be, when a cow was sick and you couldn't afford the four bucks for the vet, the cow wasn't worth it, you'd shoot it and take it into the woods. The bears would clean it up in no time. You can't do that anymore—no bears." Joe went on to relate a tale of walking down a long log and meeting a bear coming up the log right toward him. Joe froze and the bear kept coming. When the bear got close enough to spot and smell the man with whom it shared the log, it started, tumbled over, and scampered back down again. Joe tells the tale with real delight, and it is obvious he misses the bears.

"Between the hounds and guys hunting with CB radios and jeeps, they haven't got a chance," Joe explained. "Although I did know of one ol' bear, he escaped into that tangle of brush up Duck Creek we call the Torture Chamber. If you've ever been in there you know what I mean. It's about the only place a bear can get peace."

Our bear-talk continued months later at a meeting. "And another thing," Joe picked up, "we could cut down on a lot of spraying of herbicides if the bears came back. They used to eat the salmonberry shoots down so they never got out of hand." Joe and his colleague, Commissioner Bob Torppa, reminisced on how they ate the shoots too as kids—the Indians called it "shinny"—but without the bears, you don't have the tender young shoots since the salmonberry goes all rank and tall, so the logging companies spray it along with the alder.

So it's not just me. In seven years, and scores of trips into the woods, I have yet to see a single black bear in the Willapa Hills: and that seems perfectly strange to me, for the hills *look* like bear country. I began to think it was bad luck, or that bears were more secretive here than where I know them elsewhere; or maybe I just wasn't looking hard enough. But no: according to the guys who should know, the old-timers, the hunters, the loggers, and the pioneers, there really aren't many bears around these days.

One of the few palpable benefits of clear-cutting is the creation of vigorous seral scrub that black bears like. They require a diversity of habitats and foods, mature woods as well as successional; clearcuts by themselves won't do. But as part of a landscape matrix, cuts and burns and blowdowns and insect infestations create that untidy adolescent kind of forest that just suits scavenging bears. So by all rights, one of the compensations of life in a largely logged-off land should be frequent sightings of bruins. But it doesn't work that way. Not satisfied with taking the trees, the timber companies have taken away the bears as well.

"Why might this be?" ask visitors. The ostensible answer is that bears eat trees, so the timber growers dislike them. And so they do. Bears do eat trees. They favor the cambium layer of young Douglas-firs and hemlocks, which they may girdle while stripping the tasty layer; or the succulent terminal shoots of younger trees. This means they like little Christmas trees as well. If they don't kill the trees outright, they may damage their growth and form. So, on the surface, it seems that the tree growers have a case against bears. But how many trees do they really hurt?

One day in July, I hiked with British visitors on the Mount Angeles trail from Hurricane Ridge, Olympic National Park. Someone spotted a bear a few hundred feet below us. As we stopped to watch, we saw two cubs emerge from nearby foliage to join their mother. She sat on her haunches and pulled the tips of little subalpine firs down into chomping range, then she and her cubs took turns eating the new growth as we might munch asparagus tips. Made it look so good that I later tried the closely bundled, soft needles of a fir-tip myself. While satisfying to bite into, they made a terribly bitter taste in my mouth. (But then beavers love aspen bark, which is even more bitter.) After this display, I never doubted that bears eat trees.

Yet—funny thing: I also noticed that there were still plenty of trees around where the bears were working. Right there in the national park, where the bears live abundant and unhunted, lo, the subalpine forest stood intact. Extensive hiking at different altitudes failed to reveal a single area where the bears could be implicated in deforestation or serious stand-damage, outside of small glades here and there. Porcupines, I am not alone in observing, take a greater collective tree toll than bears. Of course, those Olympian forests were not the cornrow conifer stands favored by the corporate timber interests. Rather, they shag and poke and clamber together at different heights, ages, and species, making mixed systems of which the bears and their appetites are a part.

The logging giants (who had names more like Weyerhaeuser and Zellerbach than Bunyan) wanted no such dynamic woodlands in place of the old growth they first removed, so they initiated practices to discourage undesirable species. One of the best known and least popular of these is aerial spraying of herbicides to prevent regrowth of alder. Another is reduction of populations of animals that might eat the preferred trees. Bears were targeted, along with concentrations of deer and elk that might browse on young conifers. Soon the belief that bears and firs could not coexist became widespread, despite their obvious coevolution.

The pogrom against offending browsers has had three prongs. First, the companies hired, until quite recently, their own professional hunters, with the remit to go forth and plug as many bears as possible. Second, private hunters are encouraged to do the same. Each year the companies print a map to the log roads of the Willapas for free distribution to hunters, on which areas of heavy game populations have been highlighted by shading. (By varying the color, they give the impression of new maps; but as far as I can tell the maps are never updated, except perhaps to add new mistakes and misleading information.) Third, the companies encourage the hunting of bears by packs of hounds in damaged areas, and the hound clubs are only too happy to oblige. Washington and several other states still permit this unsporting, medieval practice: the bears (or pumas) don't stand a chance; having been treed by the hounds, they are followed by rabid men with high-powered rifles.

These practices can continue only through the complicity of the Washington Department of Game. Without it, bears could not legally be hounded, professionally hunted, or otherwise overhunted here or elsewhere in the state. This really came home to me one day early this year when I entered Appelo's general store in Gray's River and saw posted some new game-season regulations: there was to be a spring bear hunt in this district. A *spring* hunt, when sows are with cubs and weak from hibernation, their most vulnerable time; in these hills—these hills, where I've never seen a bear at all! And with no bag limit at all.

Archers as well as riflemen and houndsmen are encouraged to hunt (in the process wounding and harassing) bears far beyond what the population will withstand. This can only be accounted for by the Game Commission's buckling under to the timber companies and hunters. Citizens' petitions to the Game Commission to stop bear hunting with hounds in certain districts have been ignored. In fact, the hound lobby has been catered to aggressively, through the adoption of "Hot Spot" black bear damage hunts. Under this program, the game department sends dog packs to the aid of private and public landowners complaining of bear damage. Packs may chase and catch bears beyond the permit boundaries, leaving nonproblem bears vulnerable. And the word is out that damage claims often originate with the hunters themselves, who tip off landowners to bogus bear problems in order to get permits. The managers of the wildlife resource go along with it, in an apparent attempt to satisfy both hound and timber interests at the same time.

So what does it get them? A few more young trees, fewer certainly than their foresters themselves will cut during pre-commercial thinning a few years down the line. It's true: when the regrowth manages to come in densely (where the soils haven't been laterized or eroded by steep-slope logging and compaction), crews will cut thousands of young firs and hemlocks so as to let the others thrive—stuffing the downed trees into the spaces between the standing crop in an unsightly manner that blocks all travel over the forest floor. What the bears took up to this point can matter little. So why not accept the modest depredations of the bears as a precursor to precommercial thinning, perhaps reducing the need for stand management later on?

The foresters reply that bears actually prefer the best trees *after* thinning: Douglas-firs twenty to thirty feet in height, twelve to fifteen inches in diameter. The Washington Forest Protection Association (WFPA) maintains that black bears destroy millions of dollars' worth of commercial Douglas-fir trees each year on public and private timberlands. Others feel that figure is highly inflated, based on extrapolated costs and profits not likely to be realized in any case. No doubt, some bear loss is incurred by the industry. But is it worth the whole of the bear population over an entire district?

One biologist engaged in bear management told me that an aerial survey for bear damage is like looking for dead blades of grass here and there in a lawn. Except for the occasional patch, where a troublesome bear could be dealt with on an individual basis if necessary, the degree of damage appears to be greatly overrated.

Bears should be viewed as a contribution to the vitality of the forest. If they entail costs to those privileged to harvest the forest resource, then the cost must be borne. Black bears should be a part of every working forest in Washington. Those glorious forests that built the industry should have all their working parts intact—that's how they were made in the first place. The greatest temperate forests on earth had bears in them. Why shouldn't their successors, pale and depleted replacement though they be, have their bears too?

Interestingly, one of the bruins' chief persecutors thinks the woods should not be bearless. Ralph Flowers is Animal Damage Control supervisor for the WFPA. As such. he has overseen the battle against bears in the woods for ten years. Prior to that, Flowers served WFPA as a forest protection agent for sixteen years, trapping and snaring bears himself. He is responsible for recommendations to the Game Commission concerning bear damage hunts. Yet, Flowers claims a strong concern "for the bear as a resource to be enjoyed by future generations."

Pointing out that Japanese foresters on Honshu have controlled bears so effectively as to have eliminated entire local populations, Flowers has been looking for another way. "A system of reducing bear damage levels other than killing bears," he says in a WFPA report, "would prove an invaluable asset to the timberland owner as well as to the wildlife resource."

Flowers explains that black bears strip the bark of rapidly growing, fifteen- to thirty-year-old conifers in order to feed on the exposed sapwood. This takes place mainly between the middle of March and the end of June, when the cambium layer is changing to sapwood and the bark peels readily. Bears cease stripping trees when the trees grow tougher and berries ripen in summer. Sugars, trace minerals, and perhaps some additional elements are thought to comprise the trees' allure for bears. Noting that damage tended to occur shortly after bears emerge from their winter dens, "when the animal's physical condition and the availability of nutritious food are both at their nadir," Flowers proposed that a partial solution might be achieved through an intensive supplemental feeding program during the critical period.

Subsequently, Flowers and the WFPA launched an experiment involving sugar-beet pellets impregnated with minerals that were left about in areas where bears might find them. The objective was to satisfy the beasts' spring needs with a ready source that would cause them to leave the trees alone. First year results on five test plots adjacent to the feeding stations were encouraging, showing substantial decrease in damage levels. Better pellets will be field tested in 1986.

The study suggests that a full-scale program of supplemental feeding, if successful, would be cost-effective. And the fact that such an approach exists at all shows the WFPA's sensitivity to public dissatisfaction about continued bear slaughter. "As Ralph nears retirement age," a WFPA press release confides, "he has been looking for a way to save the bears as well as the trees." I would guess his retirement would be more satisfying if he succeeds. And I, for one, would be happier.

Of course, there is no guarantee that the timber companies would adopt the bear-pellet procedure across their empires. The Willapa Hills, while included in the study, will apparently provide "control" sites where spring hunts and "Hot Spot" hunts will carry on as usual for the time being. While I am intrigued and encouraged, I plan to withhold unbridled praise for such a reform until I see it in practice throughout the region.

Some may object to the unnatural nature of the feeding program. But this in itself would be nothing new: since when have the timber growers followed any natural precepts whatever? Whatever nature does, they can do better, they think. It is not that one opposes management;

even nature reserves require intensive management at times. But land management should be, in Ian McHarg's fine phrase, design with nature—not instead of nature. A highly manipulated forest yields—for a limited time, until disease or erosion or overcutting or market factors take over—a higher profit. So nature gets dumped in the balance, along with the bears. At least this new form of manipulation promises to redress that balance to some extent—and I'm all for it.

Ironically, just as the pressure from the logging lobby should be cooling with the retreat of the corporations to their big-city lairs, the hound clubs are growing in power and organization. A wildlife agent I spoke with figures the dog-followers exert more pressure to keep the heat on bears than the timber companies do.

The combined result of all these forms of directed persecution, together with the removal of nearly all the big-timber sanctuaries, hibernation sites, and privacy by overextensive logging, has been the extreme depletion of the black bear in the Willapa Hills. Only on Long Island do they remain numerous, thanks to isolation and old growth. We have lost a grand wildlife spectacle to the profit books of the timber bosses and the bizarre pleasures of the hound hunters; all the people who care about bears—many loggers and hunters included—have lost out in the process.

But that's not all of the story. I frankly don't believe that the bears make that much difference to the profitability of the logging sides. The extensive experience of many woodsmen with whom I have talked makes that clear. There is another factor operating here, and I believe it has to do with historic attitudes toward predators in general, as well as a universal need for a scapegoat.

Dr. Stephen Kellert, of Yale University's School of Forestry and Environmental Studies, has conducted extensive polls on American attitudes toward wildlife. His fascinating findings have shown, among other things, that predators continue to receive the brunt of public censure. True, attitudes have changed sufficiently for bounties to be mostly a thing of the past and for poisons to have been reduced in predator control (although the

Reagan administration has brought back the use of poisons on the public ranges banned since Richard Nixon's term). Wolves and mountain lions now have their vociferous clienteles, just as whales do, calling for their conservation and celebration. But as Kellert has shown, this constituency exists largely among the young, well-educated, urban population. Rural attitudes still tend to debase predators and to consider them enemies to be tightly controlled, if not eradicated.

Of course, there has been some historical cause for competitive feelings toward predators; these animals did pose a certain threat to humans and livestock. Yet incidence of wolf and mountain lion injuries to humans has been vanishingly small; and while predation on stock has always occurred, it can usually be controlled by removing the offending individual animal rather than attacking the population outright. Even if staunch attitudes toward predators may have had some basis in reason during pioneer days, they no longer make any sense.

Nonetheless, wool growers, cattlemen, hunters, foresters, farmers, and others continue to clamor for the wholesale destruction of coyotes and bears and the strict cordoning of cougars and wolves. And the people in general, especially rural people, go along with it. They may not string owls and hawks up to the fence posts the way they used to do, but they still view raptors and predators in general in judgmental, suspicious, and often downright negative terms. Of course, there are many exceptions, and predator enlightenment is slowly spreading across the land. But the commercial interests seek to keep the big-bad-wolf myths alive.

How does the antipredator lobby get away with it, when the facts are against it? Why do the people give it the time of day? There are reasons. For one, many of the rural people I know are hunters, so they don't oppose the killing of animals in general, and more than likely they've shot a coyote or a bear themselves. For another, they suspect the urban-environmentalist/university establishment that proffers the new attitudes and sympathize instead with the agricultural interests on the other side of the predator issue. After all, their meal ticket comes off the land. If they can be convinced that predators threaten their livelihood, they will be willing to buy the kill'em-off line. Fortunately, a lot of farmers and foresters are smarter than that.

But one other aspect, I believe, keeps the lies alive: predators make a great scapegoat. Working on suspicious attitudes to begin with, it's not hard for problem-fraught producers to enlist sympathy for their contention that predators are the roots of evil. Wool growers in Wyoming, for example, experience high mortality among their lambs due to rough terrain, harsh weather, disease, drought, an almost total lack of postnatal care, and other factors, among which coyotes come way down the list. The numbers are in on this, and it's clear that coyotes scavenge on dead lambs a lot more often than they kill live ones. So the sheepmen, in their frustration at trying to run sheep where they don't belong, where they have converted enormous tracts of public land from rich shortgrass prairie into poor sageland through overgrazing, at the expense of the pronghorn as well as the range, heavily subsidized all the while, need a scapegoat: and the coyote just fills the bill, along with (absurdly) eagles. The poison pogrom across the West, deadly to nontarget wildlife as well, has been the result.

Likewise, cattlemen losing stock in dangerous terrain are likely to blame cougars, wolves, and grizzlies first and ask questions later. So it is, I believe, with foresters and bears: while some loss due to black bears no doubt occurs (just as some cougars eat cows), they receive the blame for a great deal more damage than they do. When a stand fails to regenerate due to damaged soils or poor reseeding or planting, it's easy to blame the bears.

So the timber interests have abetted, encouraged, and even carried out the exorcism of the bears. For example, as a part of the much-vaunted Shelton Unit project, whereby the troubled Simpson Timber Company and the U.S. Forest Service pooled their lands for long-term productive management in the 1940s, bears were persecuted vigorously. With the active cooperation of both the Washington State Game Department and the U.S. Fish and Wildlife Service, bears were hunted, hounded, and trapped into rarity. Around five hundred bears were killed in three years on the unit.

McCleary, Washington, a small town situated near the Shelton Unit, holds an annual bear festival in midsummer. I heard years ago that the Game Department opened the season and bag prior to the event so that as many terrorist bears might be brought to justice as possible; then a huge bear stew was concocted and the local populace indulged in some atavistic rite that might have come from the pages of Richard Adams' *Shardik*, having to do with the vanquishment of the beast and the courtship of fecundity.

This version seemed overblown, so I went to McCleary recently to inquire after the bear festival. It is indeed still held in midsummer each year, and it serves the function of a popular village fete. Yes, they still serve bear stew. But, as I learned from a pioneer whose family has long been involved, most of the meat now is beef. A bit of canned bear from the previous season goes into the pot for authenticity's sake. And the history of the festival had to do with two newspaper editors vying over who had the tastiest bears in their district, rather than a mass rite of revenge left over from a prehistoric age.

I saw no Neanderthals on that visit, although friends who have attended the festival have reported some on the other end of a Bud. The people I talked with, all enthusiastic about the festival, could hardly be thought of as bloodthirsty bear-baiters. My informant, a gentle ranchwoman, told me that bears were seldom sighted any longer near McCleary. "I never touch the stew myself," she added. I asked if she didn't like bear meat. "It's not that," she explained. "I just don't believe in killing the bears!"

I'm glad to know that the McCleary bear festival is not the grotesque and degenerate exercise I'd pictured, involving the slaughter of all available bears and offering up their remains in the main-street blood-feast. Still, the Shelton Unit bear offensive and others like it have accomplished the same result. And few of her neighbors likely share that sensitive woman's attitude toward bears. The pioneer prejudice toward bears and all predators is alive and well in western Washington.

My ire on the issue first flared when I read a Seattle newspaper article in the late 1960s hailing an Aberdeen woman for her singular success at shooting bobcats. As I recall, she would go out in the woods with a Luger pistol, track the cats, and plug 'em for no other reason than putative "sport" and dislike—she didn't even use the pelts—averaging around one hundred bobcats per annum! That was after bounties but before closed seasons, and she could shoot all she could find. The heroic terms in which she was portrayed angered me as much as the facts themselves. That this hateful woman could remove an order of magnitude more bobcats from the woods every year than most of us would ever see in our lives struck me as preposterous; that she could be lionized for it, outrageous.

I wrote a rejoinder, "A Case for Predators," that the *Seattle Times* published in the Sunday magazine along with a color illustration of a coyote in the sunrise by Bernard Martin. Naively, I thought my counterthrust and the writings of other champions of predators in the embryonic green press would turn the tide of predator prejudice. To an extent, attitudes *have* changed in the past twenty years; but as Kellert's survey shows, not by anywhere near 180 degrees.

As I've mentioned, one needs to recognize how deep-seated was the pioneers' antipathy toward large carnivores. Settlers had always battled the beasts, that's how it was—wolves and bears were wiped out in Great Britain soon after the Norman Conquest, California's golden bear (a race of the grizzly) went out with the gold rush and became extinct in this century; so what's new? Charles Lathrop Pack's progressive text, *The School Book of Forestry*, hadn't a good word for predators in 1922: "Predatory animals, such as wolves, bears, mountain lions, coyotes and bobcats also live in the forest. They kill much livestock each year in the mountain regions of the. Western States and they also prey on some species of bird life. The Federal and some State governments now employ professional hunters to trap and shoot these marauders. Each year the hunters kill thousands of predatory animals, thus saving the farmers and cattle and sheep owners many thousands of dollars." I remember the ads in the back of *Boy's Life* in the fifties: "Enjoy a glamorous career—become a government hunter—write the North American School of Conservation," went the line, accompanied by a Smokey-hatted, rugged character holding a massive puma's paw in his own, his burnished badge condoning the murder and promising the fun of a lifelong cat-track. The romance of tracking marauders with a carbine and a badge and that hat temporarily overcame my love of wildlife and I wrote away for the details. I believe it was a Disney film of a treed cougar being shot that made me change sides. When I finally got my Smokey hat it was in the capacity of national park ranger-naturalist rather than cougar-basher. Anyway, government hunters went out about then and the ads are no more. Yet the killing continues, hindered but not prevented by closed seasons and bag limits.

Even the Christ-figure of conservation. Aldo Leopold, advocated the extermination of bobcats in New Mexican forests in his early days as a wildlife agent. Later he changed his mind as he came to realize the importance and grandeur of predators in the ecosystem. And it was his classic study of mountain lions and deer on the Kaibab Plateau of the Grand Canyon country that really launched the popular idea of a desirable balance between predators and prey.

Ecologists have since challenged Aldo Leopold's blithe connection between numbers of lions and the health of the deer herd. When the lions are gone, the deer overpopulate and starve, went the simple argument. Now it seems the story is more complicated than that, and that predators may not "control" the numbers of prey in quite such a direct manner. Nonetheless, it is clear that predators and their prey relate intimately and disrupting one will affect the other. As this idea penetrated the collective skull, people began to ask whether we shouldn't keep a few predators on hand, for purposes of balance and as an appreciative resource in their own right. Aldo's son, Professor

A. Starker Leopold, was appointed to chair a federal commission charged with looking into and ultimately reforming federal practices of predator control.

Soon bounties disappeared and seasons were introduced for most carnivores. Yet absurdities remain: if the Aberdeen cat-killer were still around, she would need only a valid hunting license to hunt bobcats, and though she would have to confine her slaughter to the open season of October through March, she could still kill as many as she wanted in that period: there is no bag limit!

She would, however, face stiffer competition, since the hound packs and trappers pursue wildcats as well as bears. Although they take perhaps 90 percent of the bobcats killed, the hounds have fewer restrictions than the trappers. The numbers of each group are growing. Widespread unemployment in timber-dependent districts has turned many out-of-work loggers into trappers. Falling back on the first resource of the explorers, these latter-day mountain men run traplines for a wide array of furbearers. Animal-rights people deplore the leg-hold trap (I agree with them: it is barbaric) and conservationists believe the annual allowable take of bobcats and lynx set by international treaty and state policy, is far too high. But at least a value has been placed on the animals beyond a bounty aimed at their elimination. Other values, such as the experience of sighting a living bobcat, have not yet been given much importance. Surely, the number of persons who would rather watch than catch a cat outnumber the trappers and houndsmen by a wide margin.

Too late, it seems, has the new interest in predators come to Willapa. Bears, pumas, bobcats, and wolves, if there ever were any, have all been exterminated or drastically reduced by settlers, trappers, hound packs, and professional and private hunters. Only coyotes proliferate. My friend Red and others believe we have quite a few more coyotes than we need. If so, their overabundance is another expression of the imbalances caused by lack of the larger meateaters. Without the competition of the larger animals, and because they are able to adapt to human environments, coyotes proliferate. if they are a problem (I'm not convinced), then it's a problem we brought on ourselves.

I did see a young mountain lion on a logging road near Deep River a couple of years ago and considered it a near miracle. Only for a few seconds did the puma appear before vanishing into the density of the alders. I have sought cougars in many places, have even found their fresh tracks outside my tent on a Vancouver Island beach. But I had never before succeeded in sighting one.

That brief moment's encounter with the rusty, loop-tailed cat mattered enormously to me. Yet it is not an experience I can reasonably look forward to repeating in these parts. Still fairly abundant in Olympic National Park, mountain lions may be hunted with hounds outside the park's protective bounds. Cougars need undisturbed conditions, and the level of activity outside national parks and wilderness areas tends to discourage them. Nor do many places offer the abundance of game necessary to ensure both the hunter's wants and the cougar's requirements. Throughout the West, the mountain lion most certainly does not receive the lion's share of the meat.

What has the local dearth of predators meant to their prey? One might expect a concomitant increase in deer and elk, spared the substantial annual take of absent pumas. But black-tailed deer are not particularly abundant here, nor have they undergone any spectacular population boom and crash as was supposed to have happened at Kaibab, post-puma. Red told me that, as a deputy, he helped recover road-killed deer in the county in 1956. In that year, eighty-six carcasses were recovered on county roads; this year there have been three. Elk, too, are much scarcer overall than previously. Red blames doe and cow seasons, formerly employed as a sop to the companies that wanted game eliminated.

Jay Brightbill, local wildlife agent, believes that the reduction of diversity through forest monoculture and the increase in the size of clearcuts (leaving less edge) has had more impact on the game than specific hunting policies. Either way, it has to do with heavy-handed forest management, and it comes home to roost in the broken branches of the local ecosystem.

As I've said, the companies give out free maps to direct hunters to sections of their holdings where they would like to see the game reduced still further. The maze of logging roads and the maps give poachers an easier assignment, and spot-lighting is common. Joe Florek and virtually all of the old-timers I speak with agree with Red in the matter of deer: there used to be more, bigger, and better animals. So the absence of predators has not had a chance to stimulate a boom in the game herds. Rather, the heavy pressure of hunters on the herds helps to prevent the return of the cougars.

Perhaps no other issue illustrates the degree to which wildlife has been perturbed by intensive human land use better than the situation at the Columbian White-tailed Deer National Wildlife Refuge. The refuge, a strip of Columbia River bottomland and islands, stretches between the two Wahkiakum towns of Cathlamet and Skamokawa. The refuge was established in 1972 to provide sanctuary for some 230 of the 350 remaining individuals of the federally designated endangered Columbian white-tailed deer. This disjunct and distinct race of the common eastern deer lost most of its habitat to land clearing for settlement and impoundments from the Columbia River dams. When Congress decided to protect and encourage the little reddish deer, the legislation called for reclamation of some of their old habitat in Wahkiakum County.

Establishment of the refuge required the displacement of a number of farming families and their homes. This gave the refuge and the U.S. Fish and Wildlife Service a bad image from the beginning: families whose forebears had worked hard to build homes and win pastures from swampland watched them go back to what they saw as worthless wasteland, and they resented it deeply. Sour feelings intensified when the service leased part of the refuge to a single farmer, not one of the originals, for a profitable sharecropping program aimed at enhancing habitat, furnishing feed for waterfowl, and generating some management monies.

Furthermore, the locals were convinced that the whitetails thrived under the old farm system, without federal intervention; and that since the refuge came in they'd gone downhill and become pests in Puget Island fields and in gardens outside the refuge. Certain biologists believed that the USFWS was managing the refuge more for waterfowl than for deer, by clearing willow and alder scrub for grassland under the pretense that the deer needed more grazing. In fact, studies confirmed that the whitetails were not in the best of health and required the scrub browse provided by the woodlots in addition to the grass of the pastures. When I came here in 1979, few good words could be heard about the refuge.

Then K-M Mountain and many surrounding second-growth forest remnants were intensively logged. This reduced cover at the same time hunting pressure was growing. A stressed herd of more than one hundred elk moved onto the refuge. They formed a spectacle that everyone enjoyed, even those who couldn't care less about the scruffy little deer. To many locals, the whitetails were damned in any case by the attention shown them by the feds, bird-watchers, and city folks. In contrast, most of the cars and trucks that stopped along the refuge roadway to view the elk herds at dusk bore local plates. At last the refuge came to have some value in the eyes of its neighbors. The rutting great bulls and strutting big cows accomplished what no amount of government diplomacy could do.

But it couldn't last. Under the Endangered Species Act, the refuge personnel were bound to manage for the listed whitetail. The state Department of Game, meanwhile, wanted their elk out where they could provide sport and meat and trophies for tag-holders—or else to open the refuge for hunting. Neighboring farmers wanted the elk reduced, and the game department agreed, since it was responsible for animal damage to private property. New studies determined that the elk were competing with the deer for necessary browse and that they were approximately thrice as numerous on the refuge as the endangered-species recovery plan would tolerate. In spite of local resistance, organized by Skamokawa resident Steve McClain and supported by county residents of all stripes, a hunt was held, and horribly bungled.

Driven from the refuge, frantic elk faced a cross fire from permit-holders on the periphery. Overanxious, underskilled hunters shot elk from the roadway, over the roadway, in sight of the massing public, on the refuge and in the water, where the wounded beasts fled in their frenzy. It was slaughter, fraught with violations. In spite of public indignation and the local game agent's best efforts to avert such a travesty, the planned hunt resulted in more bad feelings, inhumanely killed elk, and, ultimately, at least one prosecution. In the wake of such wanton violence and lack of sportsmanship and control, an outcry arose from the throats of hunters and nonhunters alike. The locals were appalled, and the state and federal bureaucrats in charge of wildlife were castigated once again.

In crisis now, officials faced a number of volatile issues. The refuge managers, who were generally well liked and respected in the community in spite of their guilt by association, faced a seemingly impossible task. To manage for the deer, they had to reduce the elk. Yet, no one wanted a repeat of the previous carnage, nor did most people want any hunting on the refuge. What they did want was elk on the refuge, which they believed the deer could like or lump. Meanwhile, the two neighboring farmers continued to complain about elk damage to their crops and fences. An expensive, long, and high electric fence (which no one liked) failed to keep the elk off their fields. The refuge's tenant farmer shot one elk in the act of damaging his crops, as state law allowed him to do, wounding it. He threatened worse if the elk remained.

Other herds, driven from the woods by clearcuts and hunters and drawn to easy pasture, invaded farmlands elsewhere in the county. Another contingent was up in arms about the propane cannon placed by state game agents to frighten the elk back into the hills and protect landowners' interests. The noisemakers didn't work, but they did a hell of a job of awakening people all night long and spoiling the quiet of the day. (I went in search of one with a maul by dark of night, but failed to find it, as it echoed so off the hills that its location could not be determined. I had intended to place its wreckage in the river, but the device's ventriloquistic tendencies defeated me.) Game officials worried that if refuge elk were transplanted to elsewhere in the county, they would simply crowd the resident herds and increase the incidence of animal trespass complaints on private property.

In short, the deer and the elk and their managers and various clienteles all faced what looked to be a hopelessly unmanageable state of affairs. A public forum was held in Cathlamet to discuss alternatives. Too savvy to chance chaos in an open hearing, the bureaucrats dispersed interested members of the public into a number of subgroups to voice their concerns.

Each group had a "facilitator" from one of the agencies involved. These discussion leaders used butcher paper and magic markers to chart feelings and ideas and to dissipate the head of pressure those feelings had gotten up. Consensus was sought on a number of possible alternatives, all of which involved reducing the elk. Each group appointed a spokesman to report its findings. As I stated on behalf of my many-minded group, consensus was impossible, except that no one wanted another hunt on or just outside of the refuge, and we favored live transfer of elk if herd reduction were deemed necessary for the deers' well-being.

The managers managed, through this clever if manipulative meeting style, to defuse an explosive situation. When the newspaper report came out in the *Eagle*, we learned that we, the public, favored Fish and Wildlife Service policies for the refuge: a fact of which we had hitherto been unaware. That, I thought, was a deft stroke. Public resource managers are becoming very sophisticated at people management, even if they can't always get their animal acts together. But no one complained very loudly, and in subsequent actions, the managers seem to have taken some of the more frequently voiced public concerns to heart.

Prior to that time, live-trapping of elk had been considered impractical because first attempts had failed. But those tries were based on bait-traps, inconsequential on a range of abundant food. Now they brought in

Bill Clark, an expert from California with experience in drive-trapping tule elk. He instructed the local staff and supervised a drive, in which thirty-eight elk were caught. These were transplanted to the northern frontiers of Wahkiakum County, where they have more or less remained, only two having come back. A second drive this year resulted in fourteen more animals shipped out, all cows and calves.

The number of elk on the refuge currently stands at between forty and forty-five, a number considered compatible with the deer. Mostly bulls with a few cows, the reduced herd includes some large racks that still provide an impressive spectacle. The powerful electric fence, it is hoped, will prevent reimmigration to the refuge, while hunting off the refuge should keep the elk population fairly low within the vicinity. Another on-refuge hunt would be an action of last resort. The fence also helps prevent field damage, and in any case the most vociferous farmer has left the district.

I spoke with Susan Saul, public use specialist for the Lower Columbia River refuges, and Gary Hagedorn, refuge manager, before completing this chapter. They both felt that the situation was now largely under control, with the interests of the endangered deer, the elk, landowners, hunters, and wildlife-watchers all catered to adequately. So it seems they've pulled it off for now. One has to hand it to them. But what a lot of trouble.

It would be nice if things were easier, if we could put back the pieces of the ecological puzzle that forestry and farming spilled across the hills and the valley floor. For example, a few more pumas could only be healthy for the blacktail herd, if the game department adjusted the deer season accordingly. And with the turndown in timber activity in the area, they may move back in. The young lion I saw near Deep River has been seen by others; perhaps it is part of the vanguard.

We may wish as much. Yet putting back the former pieces may only complicate the puzzle today. Let one cow be taken by a crippled cougar and the stockmen will be up in arms. Let the pumas proliferate, and the hounds will be Johnnies-on-the-spot. Let the hunters get wind of a significant catamount take of deer or elk, and a few of them will demand the predators' removal, begrudging their share, as some Alaskans begrudge the wolves their rightful portion of the moose. There is no easy way, when the pattern is as mixed up as it has been in Willapa. Legatees of the logged-off land, large mammals adapt to massive impacts on their habitats as best they can. Our many demands don't make it any easier.

Willapa may be far too hacked about ever to recover its former balances. The elk and the deer and the hunters and the farmers may all remain in combat with one another, nature, the state, and the feds. And the puma may remain a once-in-a-lifetime miracle sight, for that quick red ghost was the only one I've ever seen—and there might not be any more to come.

But it's not too late for the bears. The people want to see bears. The old-timers miss them and the younger ones never had the chance. There is simply no reason to exclude them any longer. The expulsion of bears by the corporations and the hound packs, with state cooperation, has amounted to a rip-off of a public possession and mismanagement of the wildlife resource.

Now, with the declining population and the pullout of the megaloggers, Crown Zellerbach and Weyerhaeuser, the time is right for bears to make a comeback in the Willapa Hills. In Wisconsin, public care for rare bears has forced the cancellation of this year's season on hunting them. We should do the same here, at least where the bears have been all but exterminated. The current game regulations list the Nemah, Palix, and Bear River game units as closed to bear hunting in 1985-86, and that is a welcome beginning. The closure should be extended to the entirety of the Willapa Hills and should remain in effect until bear populations have regained a suitable size. After that, we should go easy on bears, so as never to lose them again. If we are to have bears in Willapa, the foresters must call off their overzealous vendetta, and the Game Commission must call off the hounds and the riflemen and the archers. The bears will take care of the rest.

A triplet, scratched in pencil into my field notebook after one day in the hills, tells the whole tale (with apologies to Stephen Sondheim and Judy Collins):

> Where are the bears?
> There ought to be bears.
> Send in the bears.

# Stump Watcher

"Birdwatching is perhaps the most trivial, nonproductive pursuit they've thought of yet." So thought the author of a 1968 newspaper column entitled "The Golden Years," who urged retirees to beware squandering their remaining time on worthless activities. I saved the article for its delightfully idiotic cheek. This spoilsport would have thought even less, I suspect, of butterfly watching, a pursuit I urge on old and young alike. Running across that old column recently, I had to laugh and wonder how the author would characterize another of my trivial, nonproductive pursuits: stump watching.

Curiously, the same box of dusty clippings embraced an essay entitled "A Good Word for Stumps" by the respected Northwest nature writer Irving Petite. He wrote of the stump farm on which he dwelt—a rough sort of ranch carved out of old forest clearcuts in the foothills of the Cascades. Most stump farmers have nothing but curses for stumps, which are to them as rocks are to Swedes. But Irving Petite has many good words for stumps. He speaks of their utility as corner posts, hutch supports, and foundations. He sings their praises as habitats for nesting wrens and grouse. Especially, he appreciates stumps as nurseries for "every new race of green-growing vegetation." Anyone who can regard a western Washington stump in a noncombative spirit knows he is right:

the moss castle of such a stump is undeniably a thing of beauty and natural utility.

However, it took Petite some time to develop his "kinship with stumps." In a chapter by that name in his book *The Best Time of the Year*, the Tiger Mountain writer explored the subject in greater detail. At first, he thought of stumps as "those monstrous, monolithic grave markers for the trees that forested this land. . . . Savagely rooted into the gravel and through the hardpan, with members that went, solidly as living tooth in jawbone, halfway to China, stumps became fitting matter for nightmares."

Well they might have, since Petite at the time was engaged in winning a farm from the stump field, and it wasn't easy. "Night and day one totes and tosses and curses," he wrote, "keeping a mental image of the pasture that will someday replace this monumentally hideous excrescence on the land." But as he found uses for his adversaries, from goat shed to livestock feeder, and as he began to see how much life surrounded a dead stump, his attitude evolved. Stumps began to take on "the aspect of a kind of ally at last," and Petite realized that they possessed "a certain black, stark, bellicose grandeur of their own." At last his "kinship for stumps" came to dominate his thoughts on the subject.

My own love affair with stumps did not begin in the greenwood of wet western Washington but in arid Colorado. The first was the butt of a venerable plains cottonwood on a North Denver street corner in my childhood neighborhood. The tree had no doubt predated Denver itself. Its stump greeted my mother and me as we walked to and from Beach Court Elementary. No plethora of green things sprouted from this dry-land hulk; nevertheless it always offered something to fascinate a kindergarten naturalist: a growth of fungi, a black beetle, perhaps a hide-and-seek chipmunk. The day they blasted that stump out of the ground and carted it off, I was inconsolable.

Another plains cottonwood took its place in my affections following a move to the suburb of Aurora. Exploring the outlying fields in their last years as fields, my big brother, Tom, and I discovered an immense hollow cottonwood along a prairie ditch. Still a living tree but profoundly hollow, this patriarch stood in line with many other cottonwoods along the canal bank. The Hollow Tree probably saved our lives during a catastrophic hailstorm. Tom herded me into its blackened, gaping bole as the hailstones pounded down. We huddled, bruised and concussed, while cattle were killed in a nearby field, their backs broken by hail the size of softballs. Our shelter that day became a sacred spot of pilgrimage; hundreds of walks led to its summer shade or winter solitude. I visited the Hollow Tree at every opportunity, even after I moved away from Colorado. And when, a quarter of a century after it saved me, the Hollow Tree was struck down by another storm, cut up, and taken away, I was again inconsolable.

As a child I had frequent dreams, reinforced by Thornton Burgess books and Peter Pan, about living in a hollow tree or in a burrow beneath a stump. (The trees, always normal in size from the outside, proved elastic within when I squeezed through the hole and set up housekeeping.) So when my mother told tales of hollowed-out fir and cedar stumps in Washington so large that settlers had indeed lived in them, I had to see such a thing. It swelled my desire to visit and live in the mossy, green Northwest.

Once in Washington I never met a stump-dweller, but I did see a number of old stump cabins preserved on homestead museums. Irving Petite wrote of such structures rendered into hunting lodges, cowsheds, and a two-story cabin set in a burned-out cedar stump and named Toad Hall. Mindful of the rain, inventive settlers extended cedar shingle roofs over the sides and installed benches all around, recycling stumps as rustic summerhouses or gazebos. Other uses I've heard of included placing water tanks atop stumps, or using their broad tops as observation decks. At least one pioneer post office was situated in a hollow cedar stump in western Washington. With imagination, it seems there was no end of ways to employ the great relics of the woods.

Tool sheds, animal shelters, and makeshift cabins were among the most common products of recycled giant stumps. But not all of the cabins were makeshift. Historic photographs suggest a degree of comfort that, while it may not have matched the dens of my dreams, probably exceeded the amenity of many another pioneer dwelling.

One postcard, for example, depicts a young couple in their first home, an immense single-story cedar stump, circa 1907. A trim cedar shake roof covers the top, a pipe indicates the presence of a stove within, and honeysuckle vines entwine around the neatly hung double-sash window. The young wife peers out of that window, bouquet in hand; and if her expression seems more resigned to than excited about the situation, her handsomely suited husband standing beside the door looks as if he finds it quite droll.

Another old photo shows an aged couple taking their ease on chairs in the dooryard of their much taller stumphouse. The gent, with long white beard and hat, and the woman, in a high-collared, long dress, both seem to be demonstrating their nonchalance about living in a stump by holding newspapers in front of them as they stare directly into the camera. Meanwhile, the younger members of the clan are looking at other sections of the paper while seated on the roof, some fifteen feet up. It seems that, had I but come to Washington a good many years ago, I would not have had to look far for the dwelling of my dreams.

Now I live in southwestern Washington. It is a land of logging, and the leftovers are stumps. We owe the stumps themselves to the kind of logging performed in the early days. Limited to axes and crosscut saws, loggers naturally sought to minimize the effort involved in bringing down the giant trees of the old-growth forest. Because their butts flared radically, these trees might be twice as great in diameter at ground level as they were six, ten, or twenty feet up the trunk. In those profligate days there seemed little to be gained by sawing through these flared trunks, which in any case were very difficult to cut. As Donald H. Clark pointed out in a 1959 article, "cedars flared more than firs, but the bases of old firs often contained so much pitch that sawing was impossible. The pitch also degraded the butt logs for use in sawmills."

Fallers preferred cutting higher up the trunk to avoid the hampering brush as well, and to display their bravado and skill. So they chopped wedge-shaped notches out of the trees, into which they thrust steel-booted planks. Standing on the springboard, a logger could attack the trunk at a more reasonable diameter of, say, ten rather than twenty feet; or he could cut another notch still higher. In this manner, many stumps came to be one or even two stories high. When fire swept the stump fields, it didn't begin to remove them but often hollowed out their rotting centers—leaving stumps ready for occupancy.

Now that stumps surround me on all sides, my fantasy of a home in a stump has largely subsided. Yet I have learned that these vertical heaps of rotting wood offer homes to a great many more pioneers than the handful of resourceful humans who managed to find shelter within their hollowed boles.

When a forest monarch falls, three things begin at once. First, the canopy opens up, allowing more sunlight to reach the lower levels near the forest floor. Second, a new habitat presents itself—the roots, sides, and cap of the fresh stump. As we've seen, its surface area can be substantial. The third event is the rapid colonization by lifeforms able to feed on the wood itself or upon others that can.

Of course, deadwood lacks the fertility of thick forest soil—at first. But in this rainfall regime, which receives in excess of one hundred inches of rain per year, rotting rapidly advances. First come fungi, certain species of which can actually digest the wood itself. As British entomologist Norman Hickin points out, for the world cellulose budget to exist in equilibrium, as much must rot as grows: fungi perform much of the breakdown. Bark beetles too number among the early arrivals. They help to loosen the bark and prepare the way for later colonists.

Bacteria and protozoans spread through the stump, working on organic materials present within the young system. Larger animals arrive as if by spontaneous generation: mites, wood lice, an array of beetles, flies, wasps, and other insects. Termites, equipped with symbionts in their gut capable of digesting cellulose, attack the wood directly. Decay of the sapwood works from the outside in, while the heartwood rots from the inside out. Wood borers and fungi continue to lead the assault.

Meanwhile, the stump begins to gather a green flora. Among the green plants quick to mount the pedestal of the stump, mosses predominate. Along with lichens they require little surface preparation before a stump becomes a suitable substrate. Their richness, the diversity of their textures and greens and forms, lends stumps much of the beauty of their overall adornment. In fact, many living trees and down logs in these woods bear dense carpets of mosses, lichens, and licorice ferns. (So long hangs *Usnea longissima*, the lime-green epiphytic lichen, and so thick the tresses of the chartreuse moss *Isothecium* that visitors think trees clothed with them must surely be in the process of strangulation. Quite the opposite may be true, as the chapter "Threads of the Green Cloth" points out.)

Next, as windblown soil gathers in the crown of the stump and mixes with the rich products of rot, the higher plants begin to come in. Hemlock seedlings, heaths, and ferns dominate many stumps, but a full flora of Willapa stumps would occupy pages.

In a biome where every growing tree serves as a platform for epiphytes, scores of species in all, down logs and stumps gather life piggyback from the moment they become exposed. For many species of saprophytes (feeders on dead organic material) and animals and plants employing deadwood as a physical home, remarkably specific conditions dictate which stumps and logs they can occupy. There is a marked succession to the flora and fauna of stumps. So in each stage of decay, each species of dead wood will harbor both a general, shared array of species and elements exclusive to it at that time.

In a world with an overall shortage of deadwood, in which people depend more and more on firewood for fuel and heat, their own narrow requirements may jeopardize specialist stump-dwellers. Less urgent and understandable factors can also devastate the denizens of the downwood. In England, entomologists work to allay the "tidying up" of ancient parks and royal woodlands, where rare insects are known to hold out in certain stumps, logs, and snags. For example, the very scarce and beautiful metallic green hoverfly, *Caliprobula speciosa*, confines its occurrence to a few beech stumps in Windsor Forest and a small number of other ancient woods. Certain special insects are known from but a single stump in England.

In his booklet *Wildlife Conservation and Dead Wood*, Alan Stubbs describes the variety of organisms dependent upon decaying timber as well as the management problems facing them. Some 974 species, according to Stubbs, are associated with decaying wood in Britain. He quotes ecologist Charles Elton, from *The Pattern of Animal Communities*: "Dying and dead wood provides one of the two or three greatest resources for animal species in a natural forest, and if fallen timber and slightly decayed trees are removed the whole system is impoverished of perhaps more than a fifth of its fauna." In Britain, "dead wood is now a scarce biological resource." While the same may be said about old-growth forests in North America, at least we are blessed with abundant deadwood for now. However, since certain organisms occur only in the decaying wood from very old trees, they may one day be confined to reserves, as in Britain.

Back in the forests of the Northwest, stump succession shows itself in the changing arrays of plants and animals. More easily observable to the nonspecialist is the very variety of species present on an old stump. It has become a game with me to match favorite stumps against one another for their plant diversity. The top entry so far tallied thirty-six species of plants, and that was achieved with only a rough knowledge of lichens and mosses on my part. One better able to assay the diversity of lower plants could log truly impressive numbers of species of plants tenanting stumps.

Invertebrate census would further inflate the totals. Yet as the English entomologists have found, the irony in the study of deadwood lies in the necessity of destroying it in order to assay its biota. I often lift (then replace) slabs of loose bark to see what lies beneath; but I have not yet had the heart to pull a stump apart in order to fully appreciate its charms. Besides, since, as Irving Petite found, it can take quite a lot of dynamite to dismantle a stump, I might need more than heart to do the job.

Great differences appear in the fauna and flora of stumps depending upon age, species, exposure, and general situation. Is there forest nearby, or does the stump stand alone in a clearcut desert? Have pileated woodpeckers chiseled their rectangular holes, removing certain insects while providing entryways for others? Has a black bear chosen this stump in which to overwinter, or a chickaree picked it for a cone midden, or has its hollow been buried by brambles, blocking all means of entrance for anything larger than a chipmunk?

Truth to tell, universes revolve within these massive remains. I travel in their systems each time I enter the gravitational field of fascination surrounding every stump. The personalities of stumps attract me as much as their biota. Curiosity and aesthetic admiration in concert make me an unrepentant stump watcher. Of course, I have my favorite stumps.

On an ocean peninsula, there stand certain washed-up stumps on salt sand whose unkempt headdresses are portable heaths of salal and huckleberry. Others in those sea places, bearing the carving marks of wind and spray, have come to resemble the totems that others of their ilk farther up the coast became at the hands of the Haidas. Every tall snag in a rainy hemlock haze becomes a painting by Emily Carr, whose oils chronicled the Indian carvings in their northern coastal setting. Each cap-sprouted piling, a water-stump in the mists and lapping river waves, evokes a serigraph by the sought-after Gray's Harbor printmaker, Elton Bennett.

Up an inland river stands a very tall cedar pillar from earliest logging days, crumbling into red chunks so that now, when wet with rain, it resembles a cracking floor of glistening bricks. Breaking out into a clearcut, one sees walking octopi, where old stumps finally rotted or burned out from beneath the roots of hemlocks, themselves lopped off since: stumps on stumps. Others sprout leafless bunches of red huckleberry stems in winter, like the thick twig brooms called besoms in England. Down in a pale alder swale, a huge red-cedar stump squats over the arch of its flared base and beneath its heavy hat of salal—red and green highlights in the lichen-painted whitewood.

One day I noted a pair of silver shadows in a forest above Willapa Bay. Sometimes, on a south-facing slope, stumps become desiccated and remain bare of growth, as their bones bleach white and naked like my Colorado cottonwoods. A wispy hemlock tossed like a threadbare feather from the crown of one; otherwise they stood above and beyond the bole-to-bole carpet of the forest flora, like spare Finnish furniture devoid of ornament.

An unusually intriguing stump caught my eye as I walked along a Columbia River beach one fall. This long-rooted, washed-up monster, like some behemoth's cast-off molar, had lain along the riverside long enough to acquire more than a score of larger plants and a respectable fauna as well. Yet the high winter water will one day wash it down to the sea, so that the next time I see it, it might be tideline flotsam decked with gooseneck barnacles and seaweeds instead of asters and beetles.

Swede Park, my own bit of land, acquired its name from a virtual arboretum of nonnative oaks, maples, birches, and other hardwoods planted by a homesick Swede a century ago. These have grown too fast for their strength in our wet, mild winters, and, as a result, several were cut before I came or have fallen since. From them a few substantial stumps have come to me. But they are young, and red oak rots even more slowly than red cedar, so their evolving worlds are embryonic. A rich moss velvet softens their sharp profiles but they are likely to resist diversification for years. I shall be watching.

Lately I have taken to visiting a particular, medium-sized Douglas-fir stump across the valley at the base of Elk Mountain. Lurking among the deep vegetation of salmon-berry and devil's club, it pounces out with news of some new bloomer or fresh colonist, arresting my walks and runs and demanding a closer look. Here is what a close look showed on a late winter's day:

Lying on the shady side of the valley, this stump of some five feet in diameter fairly drips with moisture: if you press its wet wood, it oozes forth that which the sky gave and the moss sucked up. Its tree had been truncated higher than my head. A thick mass of club moss clambers over much of the face, its yellow-green contrasting with the forest-green of licorice ferns cascading from wooden cliffs and ledges. Tall sword ferns project from between the prongs of the roots. Bracken lies in rusty, broken clumps around the base, where emerald fiddleheads will soon uncoil. Great ranks of British soldiers storm the stump-front, red-capped lichens getting ready to loose their arsenals of spores; while droopy, furry rosettes of foxglove peaceably seek roothold in damp crannies.

Whole seedstores of herbs peek from protective niches: spring beauty, oxalis shamrocks, tiny mints and cresses and yellow violets, others I don't recognize yet. Woody rootlings tussle for sunspace, among them salmonberry with its magenta blossoms just now opening, cascara, and wild cherry with ruby leaf buds still closed tightly. Half a dozen species each, at least, of mosses and lichens and fungi array themselves across the surface—an elaborate carpet woven and sheared so as to change pile and style at intervals. Among these I am sure, if one could but see them, must be many of those most captivating of moss-dwellers, the tardigrades or water bears. And how many other forms of microfauna, I would not hazard to guess.

If a collecting party from another planet sought to bring back a single sample most reflective of our ecosystem, they could do worse than to select a fine stump. From it they would be able to culture a respectable slice of the Pacific coastal pie of life.

Small seedlings of western hemlock, Douglas-fir, and Sitka spruce have each selected this stump as a launching pad. Which, if any, will survive? As Irving Petite wrote, "The true function of stumps is new forests and, eventually, more stumps." I wonder if any of these startlings will become trees and, in their time, stumps.

The wood crumbles in my hand, takes one more step toward soil. A spider's funnel web catches the frost-melt dew, shines like the opalescent trails of snails and slugs that wander the wood by night. Aplodontia galleries undercut the roots where snails hide in daylight, the burrows of the primitive mammals threatening to bring down the whole packed house. Barring such a calamity, this stump and all the others will make their ways back to the earth's vegetable mold in their own good time: hemlocks first, next spruce, then firs, finally cedars.

As stumps and other deadwood decompose, they refresh the soil. Beyond homes and livings for organisms, stumps and logs fulfill an extremely important role in accumulating, then passing on, forest nutrients. Forest ecologist Chris Maser has found that as much nitrogen accumulates in decaying downwood as in the forest floor itself. Calcium and magnesium are among the other elements present in dead, decaying trees. It may take as long as four hundred years for an old-growth Douglas-fir to fully decompose. Throughout that time, the tree slowly contributes its nutrients to the soil, after the fashion of a time-release capsule.

A common practice in the woods is the burning of slash. Slash may total one hundred tons per acre in some areas. While Crown Zellerbach and some other companies eschew the burning of slash, the Washington Department of Natural Resources (DNR), the U.S. Forest Service, and other timber holders routinely burn slash during suitable weather—not too dry, not too wet. This usually means the absolute ruination of otherwise beautiful autumn days over very wide parts of western Washington, from the drifting clouds of smoke. In a land where clear days are a precious rarity, many sunlovers intensely resent slash burning. So do asthmatics and others with respiratory difficulties.

The DNR insists that prescribed slash burning eliminates serious fire hazards and improves forestlands for replanting. They explain the heavy slash harbors insects and disease and makes the land ugly and difficult to navigate. There is some truth to all of this. However, many forest fires originate in poorly managed slash burns. The amenity of cleared slopes hardly makes up for the theft of clear skies and clean air. And perhaps most important, burning robs the land of the nutrients locked up in slash and deadwood on the ground. Many scientists and observers feel that the mass extraction of timber, followed by the removal of slash through burning, may short-circuit the nutrient cycling patterns on which the forest ecosystem depends—and bring about the impoverishment of forest soils.

So what has all this to do with stumps? Much of the short-cycle second- and third-growth timber, the stuff of which slash is born, has risen up around the old stumps. True enough, many of those venerable boles withstood fire the first time around, or natural fires before or since, and then show the marks and gain character from them. The great hollow stumps, after all, got that way partly through the agency of fire. But the hotter temperatures of the fuel-packed, latter-day slash burns often finish off the job, leaving cinders in place of stumps. Or at the least, they destroy the intricate living communities of plants and animals that have arisen on the stumps since the last fires swept through. Smoky skies and ravaged stumps: no fan of slash burns, I.

Of the stumps that survive the fires, most, I suspect, will outlast me. Yet they will not last forever, and when they have all rotted, there will be no more like them. Today, trees are not allowed to get that big before being cut. Besides, they are sawed off lower, then burnt. When the manmade moss towers have all become duff, only the natural stumps that come near the end of the cycle from tree into snag into stump into soil will fill the bill for lovers of stumps. They will be scarce, as rare as the old-growth trees whence they come; and rarer, since such trees live a long, long time.

So this world of abundant stumps, this galaxy of small wooden worlds, is bound to be a temporary affair. And that is sad, because a good stump is a wonderful thing with much to teach the watchful naturalist. That makes one more good reason for setting aside the last of the old pristine forest now, so there will be at least a continuing supply of snags, hollow trees, and nurse logs, enough for woodpeckers, weevils, and watchers for as long as they themselves last.

In a previous chapter I wrote of Hendrickson Canyon, the old-growth forest remnant we hope to save near here. Its chief value lies in being a living, dying, regenerating forest: the kind the timber companies derogate as being "over-mature." We need a few such places, no matter how much we enjoy the stump fields that came our way in the wake of the rest of the old growth. For all the stumps in Willapa will never substitute for the uncut stumpage when it is all gone. Nor can stumps vindicate rapacious logging practices, only improve the leavings. Yet I admit to another motive in my hopes for Hendrickson Canyon: some of those massive and ancient trees will make great stumps when their time comes to topple and rot. I'll never see them, but it may mean that some stump watcher of the future will be spared a trip to a national park to indulge his "worthless" pursuit.

I have often wondered how many people really notice the stumps. Does anyone else nod at the leaning, huckleberry-hatted, cattle-sanded cedar stump in the pasture beside the highway on the way up K-M Mountain? Or do these remnants melt into the background, part of the anonymous furniture of the landscape, like so many fire hydrants?

Perhaps it's just as well. There was a time when a lot of notice began to be taken of stumps, when many of them were split apart for cedar shakes and shingles. Maybe that's not profitable anymore, or maybe rot at last has closed the window of opportunity for cedar salvage from the old high-topped stumps. Then too, as "natural" landscaping became popular in the fifties and sixties, people began to raid the woods for smaller stumps to take home as garden "features." At least the house-sized stumps seem immune from that fate.

No one lives in stumps anymore, to my knowledge. But occasionally one still sees a fortuitously situated stump pressed into service for this or that. At the crest of a steep hill on a logging road, I noticed a solid, straight four-foot stump with a short cap of salal serving as a signpost for logging trucks. Painted on its face in blue were the words "slow—downgrade" and an arrow pointing toward the perilous hill. Well, stumps can't help the uses to which they are put, even if it's in aid of transporting their ilk to the mill. From the look of the grade, that's one stump that had better be noticed.

At least one Washington stump has been noticed by millions of people—a gargantuan western red cedar base, whose convoluted hulk must measure twenty feet across and twenty high. It reminds me closely of the biggest stumps in the world, those of giant sequoias, with which I became intimate as a ranger in Sequoia National Park. Like some of the redwoods too, this western Washington tree has been cut so as to permit cars to drive through it. Originating in the long-gone cedar forests of Snohomish County, the stump was moved to a site near Marysville beside U.S. 99. There it stood when I first saw it, and indeed drove through it, twenty years ago.

Then came the Interstate and a series of custodial moves for the trunk. I feared it had been lost, until recently when I happened to pull into a highway rest area on my way to Vancouver, B.C. There was the Marysville stump!—on a new pedestal, its oft-dismantled pieces nicely fitted together and replete with historical interpretation. No longer may one drive through it, but better, you can walk through and marvel at its sheer enormity. This one stump, at least, will continue to be seen, and admired, by the masses of travelers who may not be, on the whole, stump watchers.

For at least one day several summers ago, an entire busload of people became official stump-watchers. They participated in "The Big Stump Tour," one of the events held to commemorate the centennial of Montesano. A forest-products town on the northern rim of Willapa, Montesano served logging camps from Vesta to Brooklyn to Cedarville. As its history revolved around the removal of the old growth, the organizers saw fit to honor trees in their festival—and stumps, as well.

Put together by Dick Moulton, Gray's Harbor County Cooperative Extension chairman, the tour aimed at showing the remains of a pioneer homestead, regenerated forests and tree farms, and, especially, some massive stumps from the old-growth forest. Resource people along on the tour included Charles Clemons, woods historian and descendant of the main logging family of the area; C. T. Riipinen, who knew the vicinity intimately as a trapper; Howard Best, a retired logger; and myself, a stump watcher.

The route followed part of an old wagon road and called at stumps of several species. Clemons, Riipinen, and Best related tales of big-tree days and shades of the area's lore, and I interpreted aspects of the natural history of stumps. Eventually we came to the Carpenter homestead, deep in the second-growth forest, far back in the dark and quiet hills. The original forest here had been logged off in the 1880s. The immense logs were hauled down to the Chehalis River, a mile below, and from there floated to the mills at Hoquiam on Gray's Harbor, I learned. Soon afterward, the homesteader brought his stock to the cleared land—a spot now deep in the shadow of conifers once again.

I found myself more eager to listen than to speak. I feel that, in concentrating on the biological, I often miss the human side of the history of the hills. More than just logging, it is a history of hardship and resourcefulness. Who now, for example, can imagine living where Old Man Carpenter held out well into this century? Remote from the smallest of settlements, with deer and stumps for neighbors, he must have been a man at home with nature.

Rapidly returning to the soil, the Carpenter homestead wrapped its furtive remains around the biggest stump on the tour. The stump tourists clambered around the knees of the old cedar, a circuitous route, or detoured right through rotted crevices and hollows. A child managed to penetrate the interior but found it spooky and scuttled out again. (I thought of my own childhood passions, and how my deathly fear of spiders and compulsion to explore hollow trees would have twisted my gut at this opportunity.)

No one attempted to climb the tower, which bore a thick headdress of shiny salal and other shrubs; just as well, as he would have damaged the sensitive fabric of plants all over its pitted sides. What, if anything, did Carpenter use the stump for? And did its nearness, as a fixture in his everyday life, give him a sense of affection for it? In any case, he left it for us to enjoy.

Way up the sides we could see the notches cut for the springboard. Dick had brought an antique springboard and double-bitted ax along, and he took portraits of those who wanted them with these implements of old-time logging—dwarfed before the stump, of course. Most of the people dispersed into the woods in search of relics of the homestead—planks from a fence, shards of crockery, metal pieces of a cookstove, and a lantern—but not before they'd had a good, hard look at the stump.

I wish our tour could have taken place a little sooner and included the cat skinner (bulldozer operator) responsible for this section of Weyerhaeuser's Clemons Tree Farm roadway. For when we made our last stop, beside a Brobdingnagian ten-by-twenty-foot spruce stump a little farther on, we found that it had been pushed over by a Caterpillar tractor employed to clear the roadside. It wasn't necessary; the operator clearly could have avoided the great survivor. Perhaps he took it on as a challenge. Whatever; he was not a stump watcher.

As the Big Stump Tour headed back for Montesano, the old-timers told tall logging tales among themselves. The town-dressed stump tourists turned their thoughts inward toward business in the lowlands. I wondered whether the day had brought anyone closer to a sense of Irving Petite's "kinship with stumps." Perhaps that's asking too much. A sophisticated relationship with stumps might take years to develop. At least, I could tell that everyone on the bus was impressed by what they saw, and I suspect they will regard stumps more acutely and sympathetically in the future. I think there should be more stump tours. These fern-draped, living pillars are definitely one of the underappreciated outdoor resources of the great Northwest.

For a naturalist living in a land of logging, compensation must be sought for the daily, depressing vista of endless clearcuts; solace for the melancholic thought of great forests that are no more. For me, one such payment comes in the form of the very things the loggers saw fit to leave behind: those mossy, rotting, wonderful stumps.

# Part IV

## Out of the Mists

# Rain-Forest Year

Nothing touches our inner circuits quite like the seasons. The rolling months and the changes they bring in our surroundings signify time passing, connote the flow of things. Whether melancholy or good cheer, fear of time flying or joy in renewal, the feelings brought on by the seasons must infect almost everyone.

Some say there are places without seasons. This is not true. In the depths of the cement cities, sycamore leaves crinkle brown, fall away, then push out again as soft green rockets that explode with summer into big fans. Crocuses thrust cheeky color-spurts from littered bits of soil. Too-cold-for-comfort trades places with too-hot-to-handle. The cities have their seasons.

So do the tropics. Quite right: the day-length varies little from January to June at the Equator. The sun shines every day; every day is hot. Few trees lose their leaves all at once, and the birds stay home in winter. But the northern warblers join them at migration's midpoint, making the passage of birds an additive seasonal feature instead of an obvious loss. Too, the monsoons come and go in the tropic zones. Travelers in equatorial regions prepare for "rainy seasons" and "dry seasons" rather than heat waves or blizzards, and dress accordingly. Seasons come to the tropics for sure, but subtly so: they creep through the tropical rain forest like a quiet night animal, instead of crashing through the countryside like the mad moose of Minnesota seasons.

With precipitation exceeding 100 inches per year, the woods of Willapa may rightly be called rain forest also. The term "temperate rain forest" more commonly refers to the coastal valleys of the Olympic Peninsula to the north, but the Coast Range qualifies by any measure of moss or rule of rain. And, as in the tropical rain forest, the seasons walk softly in Willapa.

Here the movie of the months rolls past in shifting shades of green, in a rotation of other-colored accents. The monsoon only slackens, but seldom goes for long, and some years never ends at all. Yet, occasional Mediterranean intervals punctuate the Nordic calendar of mists. They always seem to take us by surprise, because these sun-days may come in any month of the year, as likely in January as June.

January 1985, for example, stayed sunny throughout. The river dropped to summer level, and the licorice fern curled with drought and frost brought on by completely clear nights. How difficult it was to stay inside and work! The temptation to go out was stiffened by the knowledge that another January might dump thirty inches of rain (as January 1986 in fact did) and that the coming month of February might be soggy, and March as well. As we watched butterflies and robins in January and listened to radio tales of record snowfalls in the Midwest, we of course felt quite smug.

This does not sound like rain forest, nor does it seem very seasonal. But I have been here long enough to see the true faces of the months behind the changing masks and moods. I know that regardless of our fickle calendar, weather remains the hottest topic in town. Why do people discuss the weather so much? I never have thought of weather talk as small talk. After all, it's what's going on around us all the time. And it is the changing weather that orchestrates the seasons, as each day deposits its unique hours of weather into the common pot we call the climate.

Weather and climate may define the seasons, but only in one sense. There are several other measures. Seasons come in colors. They find form in the progressions of plants, from bud to leaf and flower to fall. The coming and going of the birds marks the seasons as surely as the length of day and the sun's journey across the firmament, which is of course the source of the seasons. The night sky changes, and the tides, and the level of rivers; the quality of light and kind of cloud; the pelage of ponies; and the sounds from the river at night. A great many details in the texture of the countryside change with the sheets of the calendar months.

I find myself watching these changes, just as I talk about the weather at the post office and the store, trying to anticipate the seasons, to feel and enjoy them fully, and perhaps hoping to hurry them along a little bit. As I watch, I write: journal entries, field notes, letters to snowbound friends, essays. Sometimes in my mind, driving on the freeway to Seattle or Portland and listening to Vivaldi's *Four Seasons*, I write fresh scenarios for the music such as the Venetian never imagined.

Always writing the seasons. Why? To live them more—to ratify the days—and (I suppose) in a desperate effort not to let them get away. All seasons pass, by definition. But a season written of, though only a part of its shortest day in one spot of its limitless realm, is a season that cannot escape clean. And to share. In this chapter, I plan to share some of the page-pictures I have drawn from fragments of the months of Willapa's year.

Some pagans think that the months are mere mechanical devices, contrivances to place births, deaths, marriages, the signing of contracts and levying of taxes, and that the year should be subdivided by other, more earthly measures. I disagree. I believe the ancients demarcated the months with close attention to their personalities. When they named May and November, they had something in mind beyond tithes and festivals. Perhaps women, linked to the months and the moon as men may never be, have always known this. The months *mean*, and each has its own qualities.

Why not begin, then, with January? Close enough to the winter solstice to make sense as a new beginning, January 1 means mostly food, drink, and football to a majority of our people. To me it signals time to start a sharp lookout for the first violets beneath the camellia and to walk the bounds of the land to see what's going on. In this spirit, let us walk through the months of the rainforest year.

January, most years, sees the falling of the heaviest rains. Then the world seems truly awash in Willapa. But it can also bring high pressure off the ocean, and with it, flirting interstices of sunshine. No matter how often they occur, the wet, gray lambasts always bring a sense of persecution, and the sun-running balmy days lend a feeling of unreasonable privilege. But whether January speaks in sunbeams or rainbows, or merely mumbles though the mossehurr, the month has a lingua franca in mist. Bright days and cloudy too usually dawn with fog. A nuisance to many, fog to me is more palliative than pall.

Many are the mornings when the whiteness of the fog floats on the whiteness of the frost, obscuring the river that mostly makes it. Ribbons of mist originate beyond Elk Mountain Ranch and spread over it in palpable threads. Wood smoke from the ranch house billows into the softer meadowmist and is lost. Not a moment passes but the mist metamorphoses, now merging into a mistbow above the barn like the lenticular cloud capping a volcano, now parting into a flying horse's tail. Yellow sun arrives on the flank of Elk Mountain as steelheaders arrive on the river. As if at the touch of the sun, the fog blanket unravels. In an instant, the frosty salad of grass becomes visible beside the bridge and the black trees that served until now to anchor the fly sheet of the fog over the fields. The bridge timbers go from cold gray to green to white-gold, and the wood smoke floats unchallenged by earthly vapors. The mist has gone.

The birds that stay behind or come from elsewhere help make January what it is. Eerily attenuated lines of tundra swans wheel into water meadows already owned by scaups. Upland, downy woodpeckers draw the ear as they pick at thimbleberry galls for grubs. Red-breasted sapsuckers swallow hollyberries, giving truth to W. H. Hudson's belief that reds show best as small bright bits in an otherwise green landscape. The normally unseen green on the backs of golden-crowned kinglets shows up against the mosses they probe for winter insects. Chickadees animate leafless scenes, decorating alders otherwise all unadorned. Great horned owls call and mate in dark night trysts, and winter wrens copulate in bloomless honeysuckle bowers by day.

Even a late, cold January cannot keep the snowdrops down. Waves of the simple white flowers appear around the old homesteads and had best be braced against. A warm January sends the outrageous yellow flags of skunk cabbage to waving over the wetlands. But their appearance, the most overt sign of short-lived winter's early demise, usually belongs to February. They are likely to flash dimly in the rain like amber warning lights in a twister, no one outdoors to see, or be forced to grow aquatic, as the waters rise. For if the sun shines in January, February may be called on to make up the difference in the budget of rain.

February brings expletives such as "rain, rain, rotterdam rain," and it often brings the flood. One February the rain knew no surcease. The roof began to leak beneath a paste of Mount St. Helens ash and oak leaves crammed in a valley. A rivulet ran beside the back door and found its way into the cellar, whose stone walls peed in a dozen spouts. One stream bubbled up out of a molehill. The ground must have been wetter than a waterlogged slug. Wet, wet, wintergreen wet. It rained 6.2 inches in Skamokawa that day, a few miles away. And the water ran deep all night long, as the sump pump and the river worked overtime.

In the morning, the valley lay underwater as Gray's River vaulted over its banks and up to the rim of the hills. The torrent covered the road and swept trees before it, yet reached no houses, all built just high enough (or long since abandoned if not). The muskrats owned most of the valley and must have wondered where to go next. And what of the moles? But a calf survived her borning in the storm, and the sun came out. A goat and a pony lay supine in the sun after rotting in the rain for weeks. In a few days the flood dropped again into discrete ponds where pintails and goldeneyes, buffleheads and hooded mergansers spattered the surface with unreasonable colors.

Some years experience no winter, mild January dripping through sodden February into spring. Other Februarys mimic May, like the one following a house fire when I really needed the sun. But what there is of cold and snow is as likely to come in February as at any time. After the California-like January of 1985, February made an abortive attempt to regain the blue skies of a high-pressure system. Then a low front came in off the sea, and the month settled into a chain of snowy, frosty, sleety, haily, and cold-rainy days. Feeding flocks of robins, varied thrushes, flickers, and fox sparrows kicked the duff of a hardwood bank to uncover insects, as tits and kinglets mined the mosses overhead. Brown came out of hiding and a semblance of winter struck.

When the full moon came around, I walked down to the river. A very cold and silent night—not even a mouse or a coyote. But what struck me was the singular appearance of the molehills. Multitudinous in the pastures, the molehills by day became white-capped volcano fields in the snow and frost. But due to the north winds and the southern sun, they remained white longest on their north sides.

This became wonderfully apparent in the moonlight, with the molehills so black against the pallid pastures, and their north faces so silver, almost nacreous, in the moonshine. I don't suppose the moles were aware.

Most Februarys rain and shine instead of snowing. When the winter clouds have been lightly shedding their wet, and then the northing sun finds a crack in the marble countertop of the sky, the valley is quick to seize the unexpected radiation. Against the kelly-green tunic of the land, the sequined cummerbund of the river glitters and wet roads loop like silver braids.

Omnipresent in the winter months, mew gulls seem to scissor the fabric with flashing blades of wings, and wheeling, shining flocks of starlings stitch it back together. Beyond, a thick green fringe surrounds every backlit fir and hemlock on Elk Mountain. The marble clouds part, and the scene's luminosity intensifies like a crumpled piece of cellophane in lamplight. It is one of those shockingly clear sunbursts that announces the return of the sun to the northern sectors.

Then, just to remind you that this is still February in Washington, the clouds close rank, all goes gray and olive drab, and the rain resumes. Banana slugs emerge from winter quarters to resume their lifelong slither and graze across the greenwood. Then one evening, the tree frogs begin in earnest their magnificat for male voices—a bunch of bassos, each with a throat in its frog.

That nightly devotion, were we allowed no other sensation, would leave us in no doubt that March was upon us. March signs its presence in other ways as well. Here, it is like March anywhere in the North: sun chases clouds and it is windy. Also a yellow time, skunk cabbage swaddling low meadows and blanketing boggy bottoms with its canary colors. My own little stream makes room for it on the narrow floor of the ravine. Here and elsewhere, on grassy banks, yellow violets offer a similar brand of tint in smaller packages. And echoes of old bounce off yellow daffodils, which may bloom anywhere—as often as not on the sites of homesteads long lonely.

Of course, the yellows come hoisted on stems of green. Freshest of all greens, the new growth of March! Soft, yellow-green leaflets of Indian plum fleck the forest with touches of new viridescence. Greens to beat the stuff of any market produce stall for sheer, succulent freshness cover the floor of the forest in the form of heart-shaped mianthemum leaves, turning the herb layer into the biggest salad bar in town. Just in time for St. Patrick's Day, the perfect shamrocks of oxalis paint the shady places with three-spot patterns of lush leaf-green.

Eruptions occur daily. Out of the twigs of trees, catkins pounce on unsuspecting scenes. Hazel's long canary danglers, alder's rufous tassels, and pussy willow's pearly tail-tips, all break the silence of the branches. Below, unexpected pushers break out of the banks of shale and orange siltstone slides. Among the limp, rotted rags of last year's horsetails, the new crop appears. The green, vegetative plants at first resemble the myriad evergreen seedlings that sprinkle the roadside before drying out and dying. The flowering stalks come out tough and dull, like some sort of tropical cycad sprouts. Out of the very same places burst the herds of coltsfoot heads. Before flowering, the coltsfoot and the horsetail rather resemble one another—vaguely malignant-looking fungoid growths entering the fresh air uninvited from the fetid blank embankments. Then the dense-packed heads of coltsfoot bloom, expanding beyond all hopes, in the surprise Roman-candle manner of all composites.

Meanwhile, up above the forest fringe, a revolution takes place in the salmonberry brakes. Unrest at swollen nodes hints of leaves to come, while riots break out at the flower buds as cerise packets of petals unfold. Casualties are high as kite-flying winds scatter a quarter or more of them all. This gives rosy overtones to the angle of repose, where blood currant reddens the rocks on its own. Together, *Rubus* and *Ribes* make a pair of shocking pinks for the spring show. Nothing blasts the news of the Northwest spring like the magenta megaphone of the currants, unless it be the carborundum gorget of the first rufous hummingbird to steal their nectar.

Nectar of many sorts, absent these three months, now becomes available in gradually growing rations. The early emergent butterflies, veined whites and spring azures, find the creamy heads of coltsfoot a godsend as they flock to it: seldom settling but somehow getting a drink. Wild cherry, yellow violets, anything going serves to tank the whites and blues and the early creamy carpetbag moths that speckle the woods all spring. But the hummingbirds have the greater thirst, and they must suffer attrition as they frantically examine every possible pot of bright color. Jean Calhoun at the post office fills her feeders daily, and still the many takers fight, sometimes to the death. By following the feeders and precocious patches of bloom they seem to get enough, but it is hard to imagine that the frantic search provides more calories than it consumes.

And when the rains run sideways through the sky, and the river rises to meet the slug-belly fog at its banks and threatens to return the rain forthwith; when the great March storms come, what then of the hummingbirds? How do the azures fare? And what do the swallows do in the rain? That's a good question. For if the winning scent of cottonwood balsam serves as the signature of spring around the ring of rivers, so the coming of the swallows stamps its official seal. One cannot help but wonder, on the harsh days, how can migration-weary, bug-hungry swallows make it?

Yet when the sun comes out, there they are—the otters of the air—violet-green streaks, gunbarrel-blue darts, investigating every hole for a home and nabbing every available insect. The hummingbirds, survivors too, rev up and prepare for courtship. A mourning cloak butterfly comes out, glides and alights, its mahogany and sapphire shine intact after winter's draining. Hibernation however has skimmed the buttermilk margins of its wings, leaving them pale and tattered. The blues too have lived out the storm, or else new ones have emerged. Along Gray's Bay, a single spring azure flits and basks on the damp clay bank, a powder-blue speck of sky fallen onto the floor of mud. Nearby, a carpet of cardamine spreads, ready to soften the tread of life and feed the larvae of the whites, and it is April.

April, rainstained but lush. Alders coming on like puffs of green smoke. You thought it was already green, and then April! The month explodes in a shower of green sparks that look like leaves. The valley pastures grow so green that they look blue, and indeed the heads of timothy and meadow foxtail do form blue crescents and swathes across the meads. If you pick through the patterns of new growth in the green ravines you find thick stands of frondose bleeding heart and its cousin, corydalis, whose pink-popsicle flowers will spatter the verdure before the month is out. The red currant hangs on and the Indian plum comes into its own after weeks of tentative unfurling: at Altoona, the image of their red versus white panicles dangling together from the river cliff is doubly striking. And stinking, for they both smell unflower-like, but that fails to deter the hungry hummingbirds.

Each year wears a different face, complected by the weather and made up with the paints of plants. In 1985 spring came a little late and cool. But when it came, it brought extra allocations of the early flowers. At their peak, Thea and I toured the back country of the Gray's River Divide. I have never seen such trilliums: each a generous double triptych of deep green and bright white, stapled together with six yellow stamens. All along the road banks, evergreen yellow violets gathered themselves into orefields, crossed and bound by the green-gold veins of running club moss.

Now the horsetails stood a foot tall. If the green plants mocked small pines, the flower stalks seemed pale poles sharply banded with chocolate and crowned with cones. Overhead, bigleaf maples branded the forest edge with their long bags of chartreuse flowers. All together they painted the maples even brighter than their winter dress of moss and ferns.

Those maple danglers attract the returning warblers, orange-crowned, yellow-rumped, and black-capped Wilson's. As much as bursting plant parts, back-homing birds spell April-ness with their calls, colors, and motion. Purple finches and pine siskins sit on dandelions, weighing them down and pinning their heads to the ground, so as to eat their seeds, perhaps the first seeds of the season. Band-tailed pigeons peck at cones and maple flowers and coo, their sickle-marked throats swelling iridescent. Cinnamon-flanked, apple-backed rufous hummingbirds buzz-bomb the watcher by the berry bushes; one large female, like an obese bumblebee, barely hovers over columbines and bluebells in the garden, perches to rest or nectar, then flops off to a vantage. Great with egg, she is almost too fat to fly.

This is pleasant to watch, but the high-fliers are the great glory of April with its perpetually unsettled weather: vees of Canada geese of course, but raptors and corvids as well, and cranes. One blustery morning at the mailbox I looked up to see a bald eagle, a turkey vulture, a Cooper's hawk, a red-tail, a pair of ravens, and a judgment of crows, all soaring and circling overhead at once, as swallows harassed the lot and hummers zinged up in vertical display trajectories. The corvids clearly played with the wind in the manner of English rooks and jackdaws.

Loath to abandon the spectacle, nevertheless I left it at that, thinking I'd had my share. But no sooner was I back at work than I heard a peculiar snorkeling sound approaching from the southeast. Vague recognition grew sharp as I bounded outside in time to see thirty-five sandhill cranes arrive overhead, circle Swede Park at treetop height, then gabble off into the northwest in their graceful, lanky way. Such are the winds of spring.

Those winds that carry the broad-winged birds can be mean to the broad-leaved trees. The rotten, cold edge of April (following fast after a perfect, hot Easter) has more than once sacked the plums and dashed the apples, grounding petals and pollinators alike. Even if they survive this most fickle of months, white flowers change the guard as May arrives. Just as the wild cherry goes over, elder flower is coming on. Soon it dots the bush-fields of the secondary woods with creampuff sprays that appear to be mirrored along the roadsides by false Solomon's seal. Serviceberry, so much more sonorously known by its botanical name *Amelanchier*, makes the occasional white splash at the edge of the forest. Pacific dogwood and madrona, still scarcer in the Willapas, brighten the banks of the Columbia only here and there. Back in the rain forest, May lights its way with pale fairy bells, queen's cup beadlilies, and the simple white blossoms of oxalis.

Just as we thought the palette must be exhausted, two new greens appear on the scene. The new growth of the conifers comes out in tufts so soft and light that it seems they could never harden into the stiff needles of spruce or the dark boughs of hemlock. At the same time, like so many yellow-green parasols opening in the spring rains, the palmate leaves of vine maple unfold from their red bracts. Pendent from a basketwork of green limbs, they overhang woodland ponds, where newts float near the surface like fat, somnolent otters. Their egg masses resemble transparent pomegranates, the dark seeds being the efts growing within.

Overhead, winter wrens give full throat to brilliant, complex songs that pierce the greenwood. White-crowned sparrows, commanding clearings, may sing their simpler song at any time of night or day. This habit makes more sense in the arctic with the midnight sun, yet some birds do it here as well. Out in the open, calving under way in earnest now, eagles scope the fields for afterbirths and scout the river for spawned-out salmon.

May, if sunny, seems the more glorious; if cold, the more regretted because May shouldn't be that way. Green deepens, and all the varied saxifrages of the woodland floor and river gorges go to flower, fringed and pale. The meadows run to gold as the first haying proceeds, and gardens grow voluptuous. At Swede Park, hummingbirds crowd the masses of rose-colored columbines that spring up in the watering-can days of May. Shrubs and perennials bloom in a paintbox of color. Pink and blue comfrey, salmon azalea, red and mauve rhododendrons, golden broom, purple lilacs and irises, the whole bound up in a dozen fresh greens and arced round with a broad ribbon of gold—the young, soft, pale leaves of the oaks that set the homestead so far apart from the evergreen forest all around.

Into June, the vegetation continues its excesses until the oak curtains go green and comfrey collapses under its own weight. Bracken uncurls its fiddleheads, reaches for new ground, and screens the edge of the wild bounds from sight. Mellow, pregnant June, with pink roses tumbling over the porch and incubating eggs about to burst open on all sides. But with the sun and summertime comes grass pollen, my particular bane, ensuring mixed reviews for the month.

June, early morning: Orioles haunt a foxglove patch, willow and western flycatchers prove true to their names in all respects. Swainson's thrush begins its fluting, and goldfinches copulate on a hickory limb. A great blue heron flicks its primaries just enough to keep from falling on me from on high, as spotted sandpipers call from the river shore, and crows cronk over the cobbles. Slugs haunt the damps, European black in the garden, native banana on the alders.

June, midday: Papilionids flock through the grounds—pale tiger and western tiger swallowtails drink deep drafts at the purple Scotch rhododendrons, Clodius parnassians clamber over hawkbit and early bramble blossom for undistilled honey. Barn swallows—swoop is a word they invented, and they do it in, out, under, and all around the covered bridge. I watch them chase one another's windowpane tails through the paneless windows of the barn, out over the rushy meadow, and up to their nest on the porch. Aloft, swifts.

June, dusk: A nighthawk over Elk Mountain. The golden fields of fresh-cut hay go a sickly yellow as horsemen forlornly clop over the bridge and a sullen gray sky conspires with Sibelius to invoke total melancholy. June can be as lonely and bleak as January. E. B. White agrees: "I don't know anything sadder than a summer's day."

June, in the evening: Skunks roam, and opossums, and raccoons. Great blue-eyed, tan-vaned Polyphemus moths petition for entry at the bedroom windows. Finding one open, they enter and flap like soft brown bats around the room. Later, the lights off, actual bats come in and out as well. In, too, come sweet scents off the rain forest; building since April, the odors of immoderate growth now coalesce with the scents of hay and honeysuckle to produce June's own heady perfume.

July is the time of spreading, shining bracken, of greens growing heavy and tired, of swimming-pool skies. Time for the cat to take shelter in the shade after a roll on the sunny stoop, as Scots broom cracks its pods in the heat of the day. Vultures and carrion beetles investigate whatever's dead and rotting in the sun. We investigate the state of berries: blackcaps and raspberries, red currants and strawberries, thimble-berries and salmonberries, blackberries and dewberries, all coming on strong. Kingfishers and otters work the river while swallows intercept insects aspiring to rise toward weasel-tail clouds.

A soft glow blanches the valley as the first porch spider of the season snags rising Jupiter in her starnet. Bats flip through the garden as moths emanate from wood and grass. Killdeer and coyotes call, and frogs, much subdued since March's crescendo, initiate their cluster croak. Clouds like a dinner platter of elvers in a spotlight rotate across the black table of the sky. The full moon and deep valley mist draw me out to be like Morris Graves's birds, wrapped round in moonbeams. But I am coated inside and out with pollen. The moon can make lunatics and save minds but has little power to stifle sneezes or sooth sinuses.

With August, the pollen passes. Hay meadows, cut a second time, lie in winnows and eddies of yellow and green. Goldfinches too take their second harvest, working the awaited thistle heads. Tiger swallowtails move over to phlox, hummingbirds to montbretia. The new nectarer on the scene, the little tawny woodland skipper, isn't picky: any nectary its proboscis can reach yields up fuel to the darting yellow jet. Days redolent of *Buddleia* recall butterfly-bush days of youth, when its bright purple spikes and cloying fragrance and those who came for its nectar were all the excitement a young boy could handle in one August afternoon. Saddle-backed cattle bring me back as they splash beneath the covered bridge, their reflections floating down the river.

Now you would not call it rain forest when you venture into the woods. In the garden, green snails feed on plums you dropped, but the forest mollusks estivate or at least hide out through the driest time. Mosses shrink, ferns curl, the sponge dries out. Bird song closes down and the greens, hacked about and blemished with insect damage, grow heavy and old. Away up in the Willapas, the fritillaries of high summer fly, seeking the violets, whose flowers have long since passed, on which to lay their eggs. The elk have gone as high as they can go, and at night the coyotes sing as low and close to the river as they can come.

The relative shutdown of August turns around in September as if everyone realizes that the season's running short and the rains will soon be here. Garter snakes and red admirals come out to bask. Apples fall, bringing deer and opossums by night, yellow jackets by day. Steller's jays and ruffed grouse catch unfallen acorns outside my study window, my cat watches voles and shrews and chipmunks and beautiful jumping mice, and a huge orb weaver over the borage tub catches honeybees that came to catch pollen and nectar while it lasts. Our one katydid species calls and Carolina grasshoppers launch their mourning-cloak wings into clattering flights. Tussock moths come out and I mistake them for russet butterflies; termites come out and I ignore them, as Bewick's wrens and Douglas squirrels lecture on whatever topic I care to imagine, probably termite control, or the curbing of cats.

In the evening, owls call—great horned, barn, screech, saw-whet. Sunny mornings, purplish coppers quarter the riverbanks. Snails abound in the early rains, and banana slugs mate on alder trunks glyphed with their castings. Second- or third-generation white butterflies follow the same foraging routes their parents did. Their pallor echoes in the silver-coin full moon rising, leaving a distillate of itself on the valley floor. With it, a heavy mist rises off the scales of the summer salmon in Gray's River. A total eclipse of the moon takes place. Crescentic bits are chewed away by earth, as if in retribution for Luna's devouring of Sol in the previous winter's solar eclipse. When the moon disappears, or at least becomes russet-pink, all the valley dogs' lunatic barking ceases—until the coyotes give tongue in a brilliant, mad paean to the celestial rarity above.

The woodlands in September offer courses in the study of maximum growth attained. Walking webby sunways, I see tresses of club moss dropping down from maples and stretching up from stumps as if they might meet, like soft green stalagmites and stalactites reaching to form a column in some inside-out cavern. Long tendrils of marah, the wild cucumber, make a green-and-white cascade off a tall hemlock. Where its vast tuber lies (another name is manroot) I can only guess, but such growing power, to scale the forest and climb back down again in a season! New Guinean is its tropic feel, abetted by the whopping great skunk cabbage leaves looking sordid in the streambeds, grown into tired, tatty elephant ears. But temperate, cool, not at all tropic in the feel of the balmy airs.

The first big storms are as likely as not to come in September. Then the wind rends the nylon clouds and hurls trees across wires, as electricity flees and autumn arrives. The coyotes sing to the absence of mercury vapor lights as happily as to the missing moon. I like the dark too.

October, and spiderwebs appear between every two possible anchors in western Washington. To be specific, the elegant orbs of the introduced European *Araneus diadematus*. A misty rain points out every one of thousands of fenceline nets with silverdrops. Many hang empty, but others have rain-deckled spiders in residence. A beauty of a web links two misty foxglove stalks, as if a mirror divided it down the middle. The bulbous body of Arachne grows fat with flies in a race to make eggs before the frost. October without webwork? Unthinkable.

Animals great and small populate October in Willapa. A puma appears above Deep River, as scores of elk arrive in the valleys below. Back home, a birch underwing moth comes to lights, and a great gray slug walks in the back door. *Catocala relicta* and *Limax maximus*, big invertebrates in black-and-white tones, may not seem impressive alongside pumas and elk; but I am impressed.

Also to lights, every October without fail, come the big yellow, heliotrope-speckled geometrid moths known variously as the autumn thorn, the notchwing, or the maple spanworm. By whatever name, their annual arrival coincides with a chill in the air, and they cannot last long—they lay their eggs, then die. Isabella moths employ a different strategy, spending the winter as larvae: woolly bear caterpillars. October days may profitably be spent watching woolly bears as they cross the roads, then vainly climb clay banks, only to tumble down every time they reach the crumbly overhang; or approaching, meeting, following, and finally turning back from the river. There will be other caterpillar winterers. On a reddening roadside cascara tree in the Bear River range, I find three species of geometrid larvae preparing for the cold—eating, stoking up, perhaps beginning to think about spinning up—two cryptic red inchworm sticks, and a green. Rich caterpillar hunting in the fall, and no license needed; but I wear red, and hope that I will not prove cryptic against the autumn colors.

Those colors furnish another reason for going outdoors in October, despite the deer hunters. I needn't go far. Swede Park, my home, is a crazy quilt of alien hardwoods, a patch of the northeast in somber-autumned western Washington. Here the sugar maples glow like torches against the still-green oaks that give away their own autumn plans with one or two branches each of precocious ruby leaves. Bright birches, a tulip tree, a catalpa, black walnuts, and sycamore maples, all arrayed around the white calendar house. They threaten to spill their several yellows into the atmosphere, turning it green. Later, a bright and poignant autumn tableau centers on a Japanese maple. Its flaming coral leaves, ranging from pale salmon to scarlet, fly off and sail, plucked by the frigid fingers of the rain. Still they densely clothe the smooth gray tree, at least until the next low blows in off the ocean; the last one stripped the sugar maples almost bare. The russet crinoline of the maple scintillates against the dusky blue of Elk Mountain beyond, mirrored by the azalea at its feet. Yellow lancets of black-walnut leaves score the crisp air as they glide on down. Molten oaken ingots fall as lightly as leaves.

Swede Park is exceptional, atypical of Willapa except inasmuch as the settlers of the region all brought much the same plants from home. But not all that glitters is exotic. Washington colors make soft autumns, but splendid. The glum hills opposite look less so for the yellow clumps of bigleaf maples that seem to have tumbled down into the cleft of the creek.

All across the hills, native maples like butter pats spatter the hemlocks with brighter color. The cottonwoods have gone gilt, and vine maples in sunny positions, together with osier dogwoods, bloody the nose of approaching November like any eastern shockers. Up the Columbia, the overall effect of oak, ash, cottonwood, and maple is that of old gold. The day ages, and the colors run to those of ale and malt whiskey, as the amber draft of autumn mingles with the liquor of the late sun.

Last year, October came in pallid tones for both native and alien trees. The sugar maples went yellow instead of orange. It was as if they knew they didn't belong in western Washington anyway, so why be garish about it? The vine maples also showed a pallor, like last year's colors faded by the sun, or daubed with the drippings of a watercolor fire engine left out to run in the rain. Why? Was the wet, cold spring followed by a long, hot summer a recipe for penumbral October? I can see why the fruits and nuts never set under such a regime. But I would think the early watering and later sunshine would favor the anthocyanins and carotenes and xanthophylls that make red, orange, and yellow when the chlorophyll flees. Chemical reasons exist for the leaves' response, no doubt. Ignorant of them, it seems to me as if they knew they'd be washed out when the rains began anyway, so why make the effort?

Of course, most of the bright things here come in shades of green. When it rains, as it will and will again, they only grow greener. But it is hard to imagine a licorice fern going garish. Green, and greener, is as it should be. No washing out of the mosses. Through the warm, bright days of most Octobers, the ferns and mosses merely wait their turn to shine.

*Goldener Oktober, Grauer November.* Headline in 1984: "November rain wouldn't let up"—over half an inch fell per day, average. A fine gray fist gripped the valley and the great god-sponge got the big squeeze. After weeks of throwing sunbolts, Thor switched to water lobs. The lily pads vanished from the pond like slowly sinking grebes and, with them, the muskrats. The rains had settled in for good. Or for a while, you never know. Still there could be scarlet days to come, when the big red oaks toss sparks and cinders from their flaming crowns, and red admirals show their vermilion bands as they drink fermenting apple juice from fallen fruit. More likely, what falls from the oaks is what falls from the sky, and it's cold and wet, not red-hot.

With November rain comes the time of the fungi. We fill our pouches with chanterelles, orange fruits of the forest floor, and watch the mushroom show. Purple russulas glisten, livid and viscid. Countless agarics spread their spores from receptacles the shape of witches' hats, capitol domes, black nipples, or distorted phalluses. Clear, white caps surround an alder trunk, growing right out of a collar of moss. Bracket fungi threaten to off-balance the trunks of trees whose heartwood they've already wrecked. And a jellied eel-like slime mold perambulates the path of a nurse log, making what progress? Perhaps more than the salamanders we find, moist animals uninclined to ramble beyond their rotted logs. Across these logs spreads a fungus called blue stain, and the turquoise fruiting bodies whose spores pass along its talent for painting wood an aquamarine tint.

In the chanterelle forest, a mile from home, on Thanksgiving, I reclined in one of the soft crotches of a mossy, massive stump, and watched unseasonal sunbeams blow sword fern fronds about. Then lay down on a moss mattress, and watched the motion of tall, slender hemlocks swaying in and out among one another in a mesmeric dance of fluid forms belying any notion of the rigidity of wood. The sky faded from periwinkle to soft slate. We returned to the valley, but I stayed out of doors, and my ears and nose took over for eyes deprived of short-day's light.

A snipe snarked on the marsh, a barn owl rasped after rodents, and coyotes spoke on all sides as I walked the valley loop. Cattle chewed and snuffled, and tardy frogs croaked quietly and slow. My nose told me that three or four kinds of wood burned in stoves, as well as oil and coal. Rain, mist, and moist earth, muck and cows, and rotting leaves all cast their scents abroad. And near a large apiary, twice I smelled honey on the cool night air.

The fragrance of the air in Gray's River, always fresh, takes on in December the acrid edge of silage and the sharp edge of chill. In its other traits, December's days and nights may come sodden or dry, blue or gray, still or stormy. But always there is a quality to the low light that is December's alone. This light may appear as a sun-smeared frosting on the grass and the valley, lying in a mercury pool beside the river. Or an oyster-shell glow emanated from the plumbeous clouds and the breasts of the mew gulls, picked up by the still roots of plants and thrown back in rain that threatens to rise to the skies. The light is short-lived and never twice the same, yet it always says "December."

The year runs down. The solstice stands by to start things up again. On the logged hills, burnt stumps roam the land on flexed leggy roots, standing still. Seedlings on their tops catch the rain in hemlock nets of needles and make plans to take over, come spring. Below in the valleys, where spruce giants once walked, snapping sheets of dunlins, gulls, plovers, and peeps swirl and wheel in separate planes and settle on the silver meadows. Everything awaits the longer days, except the rain.

In December, rampaging rot overruns the rainforest floor, making fodder for the summer's slugs to come. Buildings long abandoned give in to the double seduction of rain and rot, as decay advances elegantly. The double agents of clambering brambles and mats of moss make sure that nothing capable of falling will stand. Some find this a depressing and terrifying time, when the thinness of walls and the narrowness of time and the slender space between safety and surrender had best not even be imagined. But to me it means the opposite. For not a drop of rain falls, nor a bit of rot goes on, that doesn't add up to growth and green: to life. In the dripping, deeping green of December, with life at rest but all around, I take true delight.

Even so: come the solstice, the barely longer days, and the new year, I am ready for it, ready for the brand-new calendar without a mark upon it; like a fresh white page, ready to be written, the year itself stretches out before me. It will be like no other year, yet like every one that's gone before. The familiar face of the days strikes a new fashion for itself each spring—never the same, unchanging, classic, yet capricious.

You cannot take the months for granted here. In the first place, they slip and slide into one another so sloppily that they sometimes seem just one long, unruly season. And then, having settled more or less into a sort of recognizable pattern, they are liable to slip back out again: December into June, August in February. Yet they manage to make their seasons, these months; seasons full of change and surprise and gentle procession.

Seasons always green; yet seasons just the same, as one can tell by watching the reliable march of minutia through the days and weeks and months that make up the face of the rainforest year.

# Countrymen and Naturalists

Countryman. It is one of those words that instantly evokes an image. Or rather, any number of images, according to one's viewpoint. A vantage-dependent word that can wring envy from a city dweller, ill-founded sense of association from a suburbanite, wry smiles from a stump farmer. A word of color and euphony that forces me to abandon temporarily my preference for genderless nouns, because to change it would damage it. ("Chairman" can't be hurt, being blunt and functional.) Countryman means whatever one imagines it to mean.

The word "countryman" has more currency in Great Britain than here. Since, from time to time, I have compared and contrasted things British with their counterparts in the climatically similar Pacific Northwest, it might be useful to consider the connotations of the term in the land where it originated. A look at the periodical literature helps to give one a sense of what "countryman" may mean.

A British quarterly by that name presents itself to anyone who is much concerned with the countryside—the farmer and the shooter, the angler and the fell-walker, the bird-watcher as much as the crofter, the squire, or the weekend refugee from the city. Clearly, the more who consider themselves "countrymen," the more who will read the journal. *The Countryman* appears in a compact, substantial format like a green paperback book on stock paper, giving a feeling of sturdy good value in every issue. Its large-format, slick-paper counterpart, *Country Life*, addresses the gentry more directly, while *The Field* finds favor with farmers and horsey types above all. Although the copy overlaps considerably among the three serials, *The Countryman* seems to strike the broadest appeal among a people devoted to the countryside perhaps more than any other.

The popularity of a magazine such as *The Countryman* makes it clear that many people identify with the country and country pursuits, regardless of their actual residence or occupation. This in spite (or perhaps because) of the fact that direct employment in agriculture has declined drastically in this century. Lump all these periodicals' readership with that of all the nature magazines in Britain—one or more devoted to every form of country-love from butterfly watching to fox hunting—and it becomes obvious that a large slice of the people takes to the country in body or spirit as often as it can. Most of them, I suspect, would consider themselves "countrymen" in more or less precise terms. While their passion may span a broad spectrum of shades, countrymen and naturalists of all stripes seem to share a lingua franca built on a common pleasure in the "green and pleasant land" that is Britain.

As I have said in other essays, things are not always the same between the U.K. and the Pacific Northwest, in spite of their manifest similarities. In this country, and particularly in the West, countrymen and naturalists tend to remain more disparate as groups, a fact reflected in their respective magazines.

True enough, Blair and Ketchum's *Country Journal* successfully speaks to both with its blend of nature-reverence and practicum. But it generates from New England, where a mature landscape and culture resonates attitudes more Anglican than just the echo of its name. Otherwise, the wide sweep of emphases and attitudes embodied in American outdoor serials, from *Field and Stream* to *Audubon*, *Grange News* to *Birdwatchers' Digest*, bespeaks no common currency of thought.

Out here, no one actually uses the word "countryman." If they did, I suspect it would be applied more circumspectly than in England. *The Oxford English Dictionary*, citing usage as far back as 1577, refers to a countryman as "one who lives in the country or rural parts and follows a rural occupation; a husbandman." This is the sense in which the word would be adopted into western parlance, where those of rural occupation still consider those from the city to have little knowledge of the country and even less business messing about with it. On the other hand, most of the people who consider themselves students of natural history, or naturalists, tend to be urbanites or relatively recent refugees from town. Must the naturalist then be a foreigner in the countryside?

A couple of hundred years ago, a large proportion of the people everywhere lived "in rural parts" and followed "rural occupations." A century ago, or even half, that category still took in the majority of Americans. Only since the Dust Bowl days has the trend away from countryhood gathered real speed. And it has been even less time since the bulk of the folk of western Washington ceased to be "of the country." Never mind that many of them were involved in purely extractive activities that could hardly be called "husbandry," still the vote resided largely in the countryside.

Now, most of the people live in the cities and the country stands largely depopulated. True enough, a great many urban Northwesterners take to the outdoors to find their recreation and satisfy their enthusiasms for wildlife, solitude, and scenery. The enormous growth of the firm Recreational Equipment Incorporated, a local company formerly known intimately as "the co-op," attests to it. But its customers tend to be day-trippers and backpackers, seekers after the ideal "getaway"; rubbernecks, scissorbills, and nature-nuts as far as many country folk are concerned—useful for their tourist dollars but otherwise rather unnecessary, not to say unwelcome.

The process of integration of country-livers and country-lovers into the new community of countrymen that one finds in Britain has not gone very far here. As often as not, the dwellers of the rural spaces and the denizens of the towns seem to have little in common. In fact, one detects suspicion and outright antipathy between today's countrymen and naturalists, sometimes. The former see themselves as the last of the pioneers and the last to really work the land for a living; they see the visitors as naive about agriculture and unhelpfully romantic in their attitudes toward wildlife and land use. They tend to suspect their liberal, city ways, their politics, and their economic power. When metropolitan legislators propose bills that will affect rural lives for the worse, their suspicions are confirmed.

The city-bred naturalists, on the other hand, frequently condescend toward country people; they either romanticize them as caricatures of rustic bumpkins or malign them as ignorant and belligerent, reactionary and insensitive toward the land they occupy while considering themselves the true lovers of that land, unfortunately destined to live and work in the city, but unable to really imagine life without the city's amenities. They fear the pallor of provincial living at the same time they envy the idealized simplicity of country ways. In general, they tend to subscribe to Ruskin's observation in 1860 that "the words 'countryman, rustic, clown, paysan, villager,' still signify a rude and untaught person."

Like all such stereotypes, these fail exquisitely in their depictions. Contradictions may be found at every turn by anyone open enough to look. As an urban naturalist grafted into a rural existence, I admit to having held some of the prejudices I described; now I hold others. Like town-gown, city-country suspicions everywhere, they tend to reflect some truth and a lot of self-justification, as well as fear of an unknown quantity.

Though mistaken, such attitudes nonetheless possess the power to stand as barriers between all kinds of country lovers and the countryside itself. Whatever such polarities mean to the people involved, they nearly always devolve to the detriment of the land in the long run. Farmland, wildlife, water, and forests need their advocates united; divided, they stand to lose these things.

Richard Mabey nicely summed up the countryman-naturalist nexus in his book of essays on landscape, *In a Green Shade*: "Although a long history of pastoral naivety has understandably produced a widespread suspicion of on-lookers' and outsiders' versions of the rural experience, this can lead in turn to another kind of bias, in which all such views are dismissed as aesthetic indulgences. Yet the cycles of birth, death, harvest and renewal which characterize the agricultural and natural worlds have been a powerful source of symbols at every level of our culture, even for those who do not live close to them."

The Willapa Hills are a place where the concept of countryman is in flux. The original inhabitants lived entirely by the land. The Chinook Indians required no modifier to distinguish them from others of an urban ilk. The first generation of European settlers were mostly the same, nearly all countrymen, nearly all living directly off the land, though often through trade with the city. People came to the wet wilderness from rocky farm or stony street because of the opportunities it afforded—to own less rocky land at the right price (free to homesteaders), to work that land (or the river or the woods), to prosper, and build new streets. No one came chiefly to watch birds.

Even the great botanists only grazed the area, then returned to St. Louis or Kew or wherever to write up and cultivate their finds. David Douglas, one of the most celebrated botanical adventurers and one whose name lives on in the tree that built a culture, visited briefly. He circumnavigated Willapa via canoe, with the assistance of Indian guides, up the Chehalis from Gray's Harbor and down the Cowlitz to the Columbia. Discouraged by incessant rain, Douglas dismissed the tedious excursion in his journal as "this unfortunate journey."

No, naturalists the settlers were not, unless incidentally. They came to work the country, to extract resources, to find a way to make the country pay their bills and feed their families. And so it continued through half of this century. "Living off the land" was no alternative-chic option, but an obligation to the choice one made. The extent to which southwest

Washingtonians depended upon the fruits of the rivers, sea, forest, and fields may be seen even today in the little towns. I especially enjoy South Bend, the county seat of Pacific County, for this reason. No other place that I know gives quite such a sense of countryside dependency as this little town, in so many obvious ways.

As you round the curve of the Willapa River that gives the town its name, you spot the glint of white mounds rising above the estuary. From a distance they remind me of the mountains of cotton I've seen beside the Kara Kum Canal in Soviet Turkmenia, with camels alongside, or hillocks of gypsum on some industrial site. But these are oyster shells—great mounds of empty oyster shells. Among other early oystermen, James Swan, pioneer writer, harvested the sweet Willapa Bay oyster a century or more ago for shipment to San Francisco. When these played out, the great biologist Trevor Kincaid and others played around with other kinds of oysters, finally introducing a Japanese species—a bigger, faster-growing bivalve that mostly replaced the natives as the chief commercial oyster. These now furnish one of the only robust industries in the area with a continuing crop.

South Bend, like Oysterville, Nahcotta, Bay Center, and other hamlets on the watery edge of Willapa, processes the oysters and leaves their remains lying around in evidence. Eventually, the bright shell heaps find their way back into the bay in bundles that have been inoculated with the oyster spat. They make the perfect substrate for growing the next crop of the tasty mollusks. In 1985, the first annual South Bend Oyster Stampede took place to commemorate the prominent place held by oysters in the town's history, heart, and stomach.

Gazing between the white pyramids of nacre and lime, you'll see the pleasing prospect of a gracious farmstead on a smooth green slope. Across the broad tidal neck of the bay, the farmhouse shines (on a sunny day) as white as the oyster spoils. Its outbuildings and near-forty spread across green-grass pastures wrested from the tanglewood. If you could see the farm's original reason for being, you'd find that it ran white also:

the rich milk of Holsteins, distilled in their stomachs and udders from the goodness of the ground.

Driving on through South Bend's main street, overlooked by wonky false fronts and the august Babylonian county courthouse, you'll pass shops and restaurants serving up seafood, much of it fresh out of the bay or the nearby sea. The trenchant scent of apple smoke floats from a smokehouse behind someone's house. The salmon being smoked might have come from the river mouth or the open ocean. Game can no longer be sold as in the old days, but the message boards of general stores speak to the fisherman and the hunters of elk and deer: "Ammo. Licenses and Punch Cards. Bait. Cold Beer." Whether by serving the city nimrods or getting in their own meat, the people here still depend upon the game.

Machine shops and an ironworks with a venerable blacksmith shop attached service the boats and trucks that take the fish from the waters, the trees off the land. And around the bend in the nonidentical twin town of Raymond, the Weyerhaeuser mills give away the major occupation of the countrymen of Willapa: forest products. Until this and other mills faltered in recent years, most people's jobs answered directly or indirectly to timber.

And so, in a couple of miles of village roadside, you see the present-day evidences of the rural occupations that fostered a young culture: oysters, dairy, game, fishing, and forestry. The present contender, tourism, makes itself known in a civic attempt to snag the traveler into town for a longer visit. I strongly suggest taking up the offer. After a pause in a main-street tavern, where a good plate of local oysters may be had for a fiver, take a stroll and a look about. Call in at the fine historical museum for a look at the old photographs of logging's glory days and walk up the hill to scan the impressive murals in the ornate courthouse dome, recently restored. Circle a couple of blocks and finish your legstretcher on the waterfront. You will probably notice a dozen signs of direct resource dependence that I haven't even mentioned.

Living off the land didn't stop with the major rural pursuits. Minor occupations loomed large in the lives of some and reveal themselves to the spotful. For many years, as you drove out of South Bend toward Raymond, if you looked left at just the right time, you saw a sign on a small wooden building that read: "G. Kirk. We Buy Ferns." I loved that sign and its moldering building, redolent as they were of time and the river rising. Here was a man, dependent upon sword ferns, at least in part.

The building tried to fall, then was pulled down, and the sign was moved to a newer warehouse some distance away. I was pleased by that. Then, for some reason, a short time later it vanished. When I telephoned the number for G. P. Kirk in the telephone book, I learned that "old Mr. Kirk," who started the business, had died, and a different evergreen company had taken over. It ships ferns to Europe, mostly, so the fern industry goes on. But with the passing of G. Kirk and his sign, one more countryman is gone from South Bend, along with the evidence of his rural occupation.

Ferns for florists furnish just one of the field occupations for resourceful types hereabouts. Drive up certain of the rural spur roads, and you'll see signs calling for cascara bark (dry only), a valuable natural ingredient of laxatives. (A bad Boy Scout joke is to give someone a weenie stick carved from a cascara twig.) A booklet boosting the towns of Pacific County early in the century promoted cascara-bark gathering as a substantial form of income. The gathering of "chittim" (the Indian name) at a price of fifteen cents per pound paid the original cost of land for many settlers. Helen Davis, composer of the state song "Washington, My Home," lives in South Bend. She tells about how her husband, Chauncey, augmented his meager teacher's salary with a bark business and thereby became the "Cascara King."

The signs also call for fir cones for seeds. The timber companies and state nursery sometimes purchase ripe, set cones en masse. Irving Petite, in *The Best Time of the Year*, described his dalliance with the cone-picking enterprise as "a free and wholesome time," the income as "clear money, however pitchy the hands became and however kneeweary the legs." Rich harvests of cones could be made where squirrels had cut and dropped them into windrows beneath the trees or made rich caches that could be robbed. The squirrels, apparently, adapted, moving over to dogwood berries or hemlock cones or upslope, then returned later. There were plenty of seeds to go around, it seems—it takes fifty thousand Douglas-fir seeds to make a pound, three hundred thousand for hemlock—as the trees are great overproducers. Petite called the seeds "the home forests' own frankincense and myrrh."

Blackberry picking used to occupy quite a few people seasonally. But with the rampant spread of abundantly bearing Himalayan blackberries across western Washington, and people more willing to go for their own, the market has shrunk to a few specialty pie shops and produce stands. Now the big crop is wild mushrooms, meaning mostly chanterelles. One of the major shipping points for chanterelles to Europe and matsutake (a honey mushroom) to Japan is situated not far north of South Bend. Madam Mushroom buys from many agents who serve as middlemen, purchasing directly from the pickers. Their signs sprout like fungi in the fall, shouting "wanted: wild mushrooms."

Far from passing into regional folklore, these and other rural enterprises have found new practitioners with the laying off of hundreds of wood-products workers in the region. Scavenging for the goods of the woods may not pay the mortgage, but it can buy beer and burgers until the benefits run out and flight becomes inevitable.

Fern picking has not (according to G. Kirk's successors) attracted many new adherents in this way. Edwin Way Teale spent "A Day with the Fern Gatherers" in the Cascades just east of Willapa in the early 1950s. He described it in *Autumn Across America*. At that time, over a billion fronds of sword fern per year were being shipped from the region. For Teale, the day meant "memories of towering trees and of the green fountains of the sword ferns extending away, seemingly without end, beneath them." But they were not without end. The heavy logging of the past thirty years has severely limited the prime fern habitat.

"We just don't have the forests anymore," the evergreen buyer told me over the phone; besides, at just fifty cents for a fifty-two fern bunch and a hundred bunches a good, long day's work, that's a lot of effort for the pay. Men accustomed to twelve or twenty dollars per hour will not easily adapt to that. Nor would the little industry absorb many more pickers.

Chanterelle picking, on the other hand, has become so popular that amateurs like myself grow even more than usually furtive about our haunts. The new professionals bag their chanterelles with a desperation and destructiveness never known by G. Kirk in the fern patch, I suspect. At fifty cents to a dollar or more per pound, chanterelles can bring a good day's wages but not without cost to the woods and the noncommercial pot-hunters. Lately the competition and impact have grown so intense that gunplay has broken out and limits and licenses have been proposed as a means of stemming the depredation.

A columnist has even suggested that forests could be hurt by removing so many chanterelles, which have a mycorrhizal association with the trees. The fungal "rootlets," known as mycelia, form mycorrhizae, or symbiotic associations, in contact with tree roots, enhancing the trees' ability to take up water and nutrients. I doubt that careful chanterelle harvest damages the forests in this way, since cutting the stipe leaves the mycelia intact in the soil, and more than two hundred species of forest fungi interact mycorrhizally with Douglas-firs. But commercial hunters pluck caps, roots, and all. The growing level of professional mushroom hunting will surely bring regulation. At the moment, it resembles the laissez-faire kind of oyster picking around the turn of the century that led to the demise of the native Willapa oyster.

The delectable golden, trumpet-shaped mushrooms may bring many dollars per pound abroad. Amateur mycologists, as numerous here as rock hounds in Arizona, consider them among the choicest wild species of edible mushrooms. These two facts will ensure conflict as long as the gourmet market expands and the only source is the wild. We may hope for cultivation, but the mycorrhizal associations between the chanterelles' mycelia and tree roots make it unlikely in the near future. Meanwhile, the number of persons taking to the woods with knives and gunnysacks (and occasionally guns) will only grow.

Chanterelle gold and fern green are not the only ready cash in the woods today. It turns out that moss gathering is on the rise as well. I had no idea that there might be such an enterprise until I saw an ad in a recent edition of the *Wahkiakum County Eagle*: "forest moss: Cash paid for Forest Moss, wet or dry, bailed or loose." And then I remembered a cryptic sign I'd seen in the hills above Skamokawa. I went to my field notebook and found the sketch I'd made of the big, hand-lettered sign:

MOSS PERMIT W71000
VIOLATORS WILL
BE PROSACUTED

I wondered at the time what this sign could possibly mean. The classified ad tipped me off. I telephoned the moss buyer, who kindly educated me on another facet of the green market. Dennis Stein runs Resources Unlimited in Cougar, Washington, in the foothills of the Cascades near Mount St. Helens. Stein operates a log-sorting yard and produces Christmas wreaths. But his primary business is moss.

The moss, chiefly one or two species, is employed by florists and nurseries for bedding material, soil cover, wire-bound figures called "topiary," a soil additive when chopped, and in various other functional and decorative ways. Stein buys the moss from the pickers and sells it to retailers. He pays about eight dollars for a dry bale of twenty-five pounds, or six dollars for the wet equivalent, amounting to sixty-four gallons. He claims that some pickers near Forks on the Olympic Peninsula can fill a thirty-two gallon bucket in five minutes and that they average $150 per day. As for Stein, he can't get enough moss to satisfy his buyers, even though picking is becoming more popular.

That being the case, I asked him about the effect on the forest. Does it come back? Stein says that proper harvesting means taking only 30 percent or so of the moss from a given log or maple bough. Then, he says, it comes back in about three years. "Besides," he explained, "a lot of the would-be pickers are kind of flaky. They don't realize how hard the work is, or they don't like to stand in the rain and get their hands wet and cold. So many of them quit after one or two attempts at picking." And many who do stick with it lease moss rights from the timber companies to protect their interest in rich moss areas—hence the sign we'd seen.

Keenly aware of the overcutting of the forest, the manager of Resources Unlimited does recognize resource limits. He sounded sincere when he said that the longterm protection of the resource base is the main thing, especially for one who lives off the land as he does. That ad, it seems, led me to a countryman of the old sort—one of the last of the true husbandmen. At least, I hope he is right about the harvesters' careful practices. The idea of money-hungry pickers stripping the rich upholstery of the forest gives me a serious chill, for one of the chief charms of the rain world lies in its treasury of moss.

If moss makes a sustainable way for a few folks to get a living off the country, then fine. I won't begrudge a little moss, if the mass of it remains. In this bleak time brought about by the serious resource-strippers, the woods and waters cannot be counted on for a traditional wage, and the need for cash is greater than ever. So as people try to adapt, they look again to the woods and the waters. An elk or a deer, cascara and clams, some ferns and fish or moss and mushrooms can help to get a family through the winter after the paychecks stop.

Many of the former niches for countrymen have run out of the hills like soil off the clearcuts in the rain. But as often happens in natural succession, new niches have opened in their wake. Some of these fit naturalists to a T and have little to do with making an economic living. It seems to me that a shift is taking place—from a populace of countrymen who came here to make a living off the land, toward one that comes for the country and will make a living in one way or another. They still harvest the old crops, chanterelles and salmon, cordwood and clams, but usually for their own consumption. And sometimes in a different way—ferns and cascara for their comity of greens and reds in the backdrop to daily life, for example. Instead of wages, they depend upon the land for interest, outlook, and peace of mind.

Of course, the old-timers have always taken pleasure from the countryside, or they wouldn't still be here in these latter years. They can appreciate the attraction. And as the newcomers prove themselves more than effete aesthetes, as they make themselves useful in the community, improve homes and clear brambles and put kids in school, they become accepted: never natives, mind you, but welcome enough. And naturalists cease to seem so strange.

This shift has only just begun here and may not go much further. But in its nascence, it anticipates an evolution of relations that has already come about somewhat in Britain: a new future for countrymen and naturalists alike, spreading the benefits of hybrid vigor upon the rundown land, in peaceful proximity.

These fresh settlers derive a different sort of sustenance. For example, our small coterie of avowed naturalists in the district includes a pair of birders on a lifetime field trip. Finding their lives in Seattle too full of Seattle to leave much room for birds, Ann Musché and Alan Richards took a sabbatical to Gray's River, and stayed. Survival has meant converting avocations into paying jobs, a little commuting, and conjuring of whole new skills. The outcome has been less secure and compensatory than the double-salary city precedent; but it has also meant being able to bird daily and to broaden from birds to butterflies, wildflowers, and many other aspects of the countryside.

Their tutors in botany have been Cathy and Ed Maxwell. Ed, a fisheries biologist, manages the Naselle Salmon Hatchery. Lacking career opportunities for herself in their remote location, Cathy decided to call upon her college major and properly botanize the Willapa Hills. Ed takes part, and with their dog, Hyla, they navigate the wormworks of logging roads or go on foot over clearcut and across ravine, combing the battered land for remnants of unsuspected habitats. In compiling the first flora of these hills, this unconventional "fishwife" has made significant and startling discoveries of plants far out of their known ranges in the Olympics or Oregon Coast Range. While Ed tends his salmon, plying one of the few remaining truly rural occupations, Cathy volunteers her time in the service of science and conservation, doing what she most enjoys.

Dick Wilson is another individual who works a somewhat traditional countryman's job and lives the life of a naturalist after hours. Trained as a paleoecologist, the former professor of biology at Southern Oregon College left academia for an oysterman's existence in the village of Bay Center on Willapa Bay. No basic oyster-picker such as formerly worked these waters, Wilson has applied the most modern techniques to the propagation of his valuable crop. Running the first oyster hatchery in the state while managing and harvesting extensive oyster beds has left Wilson little enough time to enjoy his avocation of natural history—for which he has far more than a hobbyist's background. But living on a prolific, unspoiled tideland, he has nature about him in his daily work; and he vows to spend more time in the hills investigating its secrets. A particular interest, the amphibians, beckons from the region's many watercourses. Another naturalist has found a niche in Willapa.

I fit the pattern as well. My tag line as an essayist in *Horticulture* for several years described me as an "urban naturalist." I wrote on urban biology, and I spent a lot of time in the parts of many cities that were, literally, seedy. But as my career in nature conservation took me farther and farther from nature, I felt a growing need to get back: back to the country where I'd never really lived for long, except in my genetic and imaginative past. So (with my co-conspirator at the time, Sally Hughes) I too threw in my city job and its paycheck and complexities, and came to Gray's River to live.

I came with the romantic yet semipractical notion of harvesting words grown out of the soil of the hills and the valleys. This book is a direct result; it is a kind of a crop, a new one for Willapa. I also came to spend time daily among nature; and that I do, if I never leave the precincts of my own house. For my wife, Thea, it is the same, except that she has lived in the country before; she was a countrywoman in the old sense, as orchardist, gardener, and goatherd, before she came to be a smallholder in Gray's River. But having spent some years back in town, her chief desire was to get out more. Now, gardening, botanizing, drawing ferns for transfer to silk screen, and filling her basket (never a gunnysack) with chanterelles, she is doing just that. Getting out: as much as anything, that is what articulates the motivation of all the new countrymen and naturalists.

The original settlers shared a spectrum of country skills among them, and these they taught one another. We are no different. Ann and Alan know the birds better than the rest of us; Ed and Cathy, the fish and the flowers; Thea and I, the fungi and the arthropods. We traffic knowledge and experiences among ourselves in a form of latter-day, intervillage commerce. As discoveries are made in still other areas, or as rare species or special phenomena turn up, the word goes around: a fine stand of old trees has come to light up the South Fork? The pink fawn lily is blooming exuberantly behind a nearby homestead? Coots (common elsewhere, inexplicably all but absent here) have appeared on Gray's Bay, or cattle egrets in the valley? The word goes around. Field trips arise. Canoes take to the rivers, trucks to the logging roads. Picnics ensue, or beers at the Oasis following a hike or a paddle. Never as often as we would like, since we are all a bit solitary, and slightly chary of the ultra-outing habits of our urban counterparts. But we do interchange, trading in the currency of nature.

There are others: among them Ralph Widrig, an erstwhile realtor in the Maritimes who came West to devote himself to the study of birds and plants in a duneland wilderness beside Willapa Bay; Vance Tartar, a noted protobiologist who abandoned academia's fetters to set up his own bogside laboratory in the woods, named Wit's End; Kathleen Sayce, a seaside plantswoman possessed of the arcane and valued knowledge of lichens and mosses; and David Hoffman, a state park manager who prefers a trying job in the country to one in the city, farther from the fishing holes.

The latter-day immigrants find ways to make a living that leave room for the abundant enjoyment of the countryside they came for: an intermittent seaman teams with an extension agent, a potter with a photographer, a pair of musicians take to the road for gigs that bankroll their rural existence. Some have indeed become husbandmen, raising bees, Scottish Highland cattle, or specialty crops. Freelancers write, edit, design, or consult from their country seats, traveling to the city only when necessary and to the post office a good deal more often. Artisan, garden, and computer skills collaborate as couples find ways to live lightly and pay the bills while reserving the time and freedom to enjoy their chosen locations.

Of course, I would be entirely wrong to imply that we latecomers were the only ones with any love or knowledge of nature hereabouts. The pioneers could not have survived the loneliness, rain, and rigors without a strong regard of their own for the beauties of the place. Often, they knew little of the nonfunctional, noncompetitive species in their midst, for why should they? But just as often they would have surprised us with their sophistication as naturalists, far from the library or the lab. James Swan, subject of Ivan Doig's penetrating *Winter Brothers*, may have been the consummate observer of everything around him. Pioneer oyster-picker, Indian agent, diarist, and descriptive writer, among many other things, Swan diligently recorded the lifeforms around him in the inner decades of the nineteenth century, as he roamed western Washington from one corner to another. Settlers did not have to call themselves naturalists in order to be concerned with nature.

H. P. Ahlberg, Swedish immigrant, had been a noted horticulturist in the Midwest before he came to Gray's River in the 1870s. Here he established what came to be known as Swede Park, now my residence, as the centerpiece of his valley homestead. He must have had a powerful feel for plants and the beauties they could give. Returning to Sweden in 1911, he never got to see the trees and shrubs he started come to full maturity. But we derive the benefit of his prolific planting of European and eastern hardwoods all around the valley. Ahlberg was a naturalist of sorts.

Little record has been left, but among the loggers, dairymen, farm wives, teachers, and others, there must have been a good few who valued and concentrated on their daily, pointblank contact with nature. Pioneer life had to be more confrontational with plants and animals, but it does not follow that none of the old-timers were nature lovers. But they left no account of it that I have seen. Willapa apparently produced no Suksdorff, the great farmer-botanist of south-central Washington; or Hopfinger, the immortal orchardist-lepidopterist of north-central Washington. Our naturalist forebears, in witness to the subtlety of the southwest corner, took their nature as a quiet, personal devotion, making no great show of it. As a result, the field is wide open to those of us who have arrived late on the scene.

All rural places produce their knots of young naturalists who plunk for suckers, bait crawdads, slither after the frogs and salamanders of the swamps and woods, or net insects for 4-H displays. I have encountered truants who couldn't spell or figure but who knew more about life at the river's edge than I do. A few will remain within the domain of Willapa, or that of natural history, but not many will likely do both. Our society offers little reinforcement for children to remain interested in nature as adolescence approaches. Sports and other socially acceptable activities nearly always win out. Hunting and fishing are the closest things to nature study that country kids are likely to carry on beyond the days of 4-H. Perhaps that will change as new role models enter the community, adults who may fish but who watch birds too.

No small town is without its woman who feeds and loves the birds and who studies them in her backyard fashion. Ours is Jean Calhoun, the postmaster. Compact of build, with short hair colored like a cloud, movements stylized after years and years of conforming to postal standards and duties, pursed mouth capable of sweetest smile or tentative appraisal or tart remark with nothing tentative about it at all, Jean has no double anywhere. Jean, as midwife to the mails (in league with Hermina the Mail Lady who brings it to our box in her much-adorned station wagon, with lights, seals, and flags), forms the necessary focus of life in a town of remote connections. For me, a creature of the mails, she enables life itself. Much of the pleasure of the daily function of the mail is not only knowing that my every postal need will be met in a personalized way, but knowing too that the service will come with a tale. Never a gossip but always the keen and discreet observer of humans and other animals, Jean delivers the stamps with a waxpaper sack, the correct change, and a story if she or her customer has time for it. If not, her ministrations can be conducted with the absolute minimum of verbiage on either side of the barred window.

The stories I remember often have to do with animals, and usually birds. They might concern a dog that, coming around night after night, lapped up the beer Jean put out as slug-bait and finally betrayed itself with a wobbly gait and a hiccup; or the cat who threatened her robins, and the many methods she employed to foil it before the volunteer firemen took it off her hands as station cat. But mostly, Jean tells me about the birds: the time when so many rufous hummingbirds came to her feeders that one aggressive male actually killed another; the every-morning flights of ravens down out of the hills and, gronking, up the river; or the early assembly of the swallows on the wires after a hot summer, when fall was expected early, and the afternoon light through the turning foliage made the post-office lobby golden.

One could compile an ornithology of Gray's River from Jean's observations, and it would not be unlike Gilbert White's letters from Selborne in the range of its curiosity and closeness of its scrutiny. It would contain a few apocryphal creatures that appeared in molt or during busy times at the postal window and a few others remote from their field-guide ranges. It would also include some common names drawn from local usage rather than any checklist. I was fascinated, for example, to hear Jean apply the name "yellowhammer" to goldfinches. It belongs to an English bird of bright yellow hue and may have come here with early settlers in the same way as the name "robin" came to be applied to a bird similar in color but very different otherwise from the English robin that Gilbert White knew in his Hampshire village.

Jean's ornithology would include a lot of downright fascinating behavior that this punctilious householder and tidy gardener has watched in great detail, that the birder from the academy may have missed altogether, not having the benefit of a long stay at a country post office with a feeder. Jean's duties occupy all her working hours; but they also keep her confined to the premises much of her off-time, so that she is not likely to miss many comers.

Now that I've got Jean watching butterflies, she holds great silkmoths for me that the utility men find around the substation; and she was the first to tell us of the record hatch of tiger swallowtails this year, which we missed while in England. Every interview, nearly every subject, and many remarks of mine, will be followed by the same summary statement: "So," rendered, however, in a great array of inflections as befit the situation. A Lower Columbia remark, born and bred at Clatskanie on the Oregon side and refined and matured in Gray's River, that "so" simply says it all. It is a small masterpiece of concision.

Jean is the soft bulb that lights the gray days in Gray's River, and may she long continue to shine. So!

What is Jean, then, if not a countrywoman? A former farm wife, a birder, a vital link in the communications of a country town. Here is the person in whom the distinction between countryman and naturalist breaks down, finally.

And so it should. We're all in the woods together, whether we have been here for generations or mere years, work at a rural occupation or an electronically aided cottage industry, prefer salamanders or steelhead salmon for company. In practical terms, the birder and the logger may still be miles apart. For one, country life may mean a chestnut-backed chickadee outside the kitchen window on an otherwise silent morning;

for the other, the chance to shatter that morning with a chainsaw with no one to complain about it. One may want planning and restrictions on land use, while the other regards such things as anathema. Yet I find I can talk with anyone here, find a common basis, and share some shred of regard for the countryside, no matter how tenuous.

The question remains whether the latter-day countrymen can fully know the country in the way the fifth-generation country laborer does, no matter how many salamanders or flowers they can identify. As a writer, I wonder whether my typing fingers can know anything of the land and waters the way the cold, wet, stiff fingers of loggers and fishermen know them so well. No amount of splitting wood or cutting back brambles, it would seem, can instill the state of mind that comes from felling tall timber or getting in a thousand bales of hay.

Another word harvester, Richard Jeffries, apparently felt this distance acutely as he sought to capture the essence of English country life in his essays and novels. According to Richard Mabey, "The closer [Jeffries] is read the more the major concern of his work appears to be an implicit questioning of what a 'country writer' is, an exploration of the relationships that are possible between a reflective outsider, marginal in all senses to the real business of the land, and the hard imperative of life in the fields."

As another country writer and reflective outsider, I share those concerns. I also wonder how much the field or forest worker, used to confronting nature for a hard-won livelihood, can empathize with the appreciative naturalist who confronts nature mostly for refreshment. Perhaps the countryman, *sensu strictu*, and the naturalist from town can never fully know one another's minds and must to that extent remain alien.

Richard Mabey goes on to ask, "To what extent is 'the countryside' not just a fact of geography and a place of work but an emblem for a whole range of social and spiritual values? And if it has such intangible assets, who are the rightful inheritors?"

Presumably the inheritors of the countryside would be all those who share the range of values of which Mabey speaks. But he writes from Britain, where (I contend) country-livers and country-lovers have melded, if raggedly, into a kind of common constituency. Here we seem to be a long way from that. The logger and the birder hold different views of the worth of the woods; their activities, preferences, and biases are often at odds. They have separate perceptions, a lack of common values, and, worse, mutual suspicions.

Yet, for all this, they share something fundamental. For all species of countrymen, at least those who stay by choice, are united through a common decision to live close to the land, to shun the city, on a daily basis; to be "first-person rural," in Noel Perrin's delightful phrase.

And that is the very basis of country life, as I see it, regardless of one's harvest, be it timber, fish, or pasture, butterflies, birds, or words. Why should countrymen all be alike or of one accord, any more than the citizens of town? Diversity is its own reward, after all. In the end, it's the love of the land alone that counts.

# And the Coyotes Will Lift a Leg

I wish I had said it first. Someone else did—who knows who?—and won't get credit for it. Spinners of clever quotations share the relative immortality of their words sometimes; platitude-makers, almost never. Even so, I still wish I'd said it first: "nature bats last."

On the other hand, maybe this isn't a platitude. The thing about a "good" platitude is that it should be self-evident. I'm not at all sure that "nature bats last" is self-evident to very many people at all. Perhaps it's just a platitude for pantheists, a byword for Earth First!ers and others who sometimes seem as willing to exclude humans from their concept of nature as most people are to neglect the other species. In which case, it misses the point altogether.

The point of a platitude, as I see it, is to preach a point to people who already know it but act as if they don't. I doubt that those who need to know that nature bats last have any clue at all as to what it means, even lack a cosmology in which it could make any sense. That renders it an esoteric idea, an impossibility for a platitude. I guess it isn't one after all.

An aphorism, then; a verbal balm, a tonic thought. It was meant, of course, as a warning, a shaken finger; but falling on mostly deaf ears as such, it recycles pretty well as a curse of revenge.

What does it mean? "Nature bats last." It means, we may be in the lead now, the natural world may seem the underdog and down in points as well. But when we've finished our act, hit the grand slam, or struck out (which may be the same thing), nature has an extra inning coming—all to herself, unopposed, unending. No one will be keeping score anymore, and guess who wins?

Let me be clear from the start. This is not a threat. I am not writing another admonishment to repent before the day of ecological reckoning. It is late in the day for us to clean up our environmental act, and I am assuming we will not. On the local scale (as in Willapa) and increasingly on the larger, the major decisions have already been made, and we can only live with them. But I'm not preaching doom—what a waste of time to preach the inevitable! We all die; all species die. The only question is, when will we pull ourselves off the respirator?

To me, nature's batting last is neither a warning nor a threat. It is a cheerfully flip recognition of a certainty. And a comforting certainty it is: imagine, the glory of the universe going on and on, free at last of the bad bet that was man on earth! When John Lennon wrote "Imagine," he could have added a verse: "Imagine there's no people." My humanism ends where we become so fond of ourselves that we cannot imagine the mortality of mankind.

But supposing, against all odds, we began to run the world right (a phrase that contains in its hubris the seeds of its own defeat)? Couldn't we then change the batting order? Wouldn't I at least want to hope for the endlessness of the human race?

Sure. But that's vain. The best we could do would be to postpone our departure. Any time on for good behavior would just amount to a stay of execution. To think we could indefinitely put off the end of the age of man by acting right toward the earth for a change is like taking up running in dissipated middle age in the hope of cheating death: it might work for a while. You can't prolong life forever, not for an organism, not for a species. But you can sure as hell hasten its demise.

This is harsh stuff, and there has been some harshness in some of the previous essays. I have criticized and taken account of what I feel to have been mistakes. An ungenerous reader could mutter "Cynic!" and close the book, so near the end. But I am not cynical about humans and the rest of nature. When I insist upon the mortality of all species, including our own, it is not an unhappy thought. And when I invoke that aphorism of uncertain category and origin "nature bats last," it is in good cheer that I do so. My outlook, ultimately, is not a pessimistic one. But then my frame of reference does not encompass human fortunes alone.

Let's look at outlook, for nature, humans, and otherwise. I have always been a short-term optimist, by nature. Whether that has a genetic element or comes from example, I cannot know. But I have always believed that more good things were likely to occur than bad. (This may have to do with my rather catholic tastes as to what constitutes "good.") It is a matter of being open to possibility and aware of serendipity's whisper. Everyone is invited to serendipity's picnic, but only a few bother to attend. Positive thinking? Are we headed toward platitudes? It's more than that. It's being willing to conspire with the physics of fate (chance, really) to harvest luck from happenstance.

Jung called coincidence "synchronicity" and it happens to us all if we are only aware. Coincidence—happening with. You must be ready to see it and do more than say "wow" when you do. To pluck a plum when you pass beneath the bough, you've got to be looking up. To catch the glisten of the green snail beneath the plum tree, you must regard the ground. To capture more good than bad, you scan the whole and, mantislike, snatch the happy moment before it springs away, out of reach.

I am not a fatalist, and when some great coincidence brings me joy I try not to say it was "meant to happen." Strings of bad "luck" do sometimes befall people, even those who watch for the good. My brother has had a lifelong run of bad breaks, more than his share, while I feel I've had more than my fair share of good ones. Stochastically (a word I learned to toss around in graduate school that means "chances are") one is as likely to be felled by lightning as lifted by the lottery. Life is a lottery. But somehow, seekers after something often seem to get better breaks than others who fail to look around. Or do they simply find more compensations?

Of course, another reason for short-term optimism lies in our ability to apply will and thought and action to effect change in our time. We can create a nature reserve and enjoy it for the rest of our lives. We can vote the bums out. We can live selectively, choosing that which we wish to experience. And there are, after all, far too many pleasures available to be able to sample them all: too many wild and intriguing places to ever visit, people to meet, birds to watch, books to read, symphonies to hear, and so on. The riches embarrass our poor ability to enjoy them. Pessimism in the shortterm is its own punishment, since it vitiates the will and makes one a pawn of circumstance.

Looking out toward the midterm, however, my attitude rotates. Beyond the here and now, a cautiously pessimistic outlook seems only reasonable and realistic. I suppose this means shifting out of my own life and into the many other lives on earth. Speaking of the world, there is no gravity; the earth sucks (whoever said this first probably wouldn't own to it). My, how it sucks these days. Admittedly the *Wahkiakum County Eagle* gives one a less jaundiced view than *The New York Times* might, but I also see the *Longview Daily News* occasionally, listen to *All Things Considered*, and watch what the cat brings in. How anyone can be honestly optimistic over the next century, regarding mankind, I cannot divine. I won't repeat the litany; it's there for all to see, who read any papers at all, or the walls, between the lines, tea leaves, sweaty palms, tarot cards, or the weather. Even the Bible seems to have it about right, somewhere toward the back (if not in the "to have dominion" part in the beginning).

Come to think of it, the Bible does get it right at both ends. Humans took dominion over the earth, now they face Armageddon. The story is rather circuitous from A to B and the cause-and-effect gets a bit mixed up, but it's all there. The sad part is that it gives people the idea that someone else is going to clean up after them. If they're not responsible for the outcome, if they're not culpable for their mess, how can people be expected to function with the future in mind?

The present and near future could get downright depressing if it weren't for nature. As John Hay more elegantly put it in his small classic, *In Defense of Nature*, "What is there to be optimistic about, especially in the face of enduring human perversity? Not a great deal that is predictable; but if enough of us are willing to walk out and meet nature instead of by-passing it, then we will at last belong. And when all is said and done, real stature comes from an attachment to the unknown."

Yes. And this brings me to the long-term, where for me optimism swings round like a major moon to again eclipse the darker view. Unreservedly, I am optimistic in the long run. Not necessarily for *Homo sapiens*, whose puny fate fails to concern the cosmos. But for nature, which is everything, the whole to which our greater allegiance belongs. And for the earth, which is all most of us shall ever directly know of the universe, finally to be freed from human bondage.

Here is where I differ from many theists. They see salvation from earthly dross in an afterlife for the soul. I see afterlife as salvation of earthly dross that is the soul. The perpetuation of my matter in crocus, coal, or comet is all I need know about the next act—that atoms continue in nature. We both see something coming that ratifies what has gone before and flenses the flesh of suffering. To them, however, heaven is full of personalities on permanent vacation; to me, heaven is a permanent vacation from personality.

This inability to face the extinction of personality serves as one of the main reasons for the rejection of evolution by some creationists. They are smart enough to see that, if life evolved, it will continue to do so, and that we (body and soul) may not survive the process. So they seek to preserve their cherished selves by pushing fairy tales—as if, by evangelizing hard enough, they could make it so!

The latest version of the creationist credo delivered to my door is a "textbook" that its makers, the Jehovah's Witnesses, hope to have placed in schools. Entitled *Life—How Did It Get Here? (By Evolution or Creation?)*, this heavily illustrated and simply written tract attempts to convince the reader that evolution is a "lie," claiming: "We should feel even stronger indignation toward the doctrine of evolution and its originator since the intent is to defraud us of eternal life."

The "marvelous new era" that this book promises for believers will have peace and plenty and endless health, youth and life for all. Apparently there will be room for endless population growth, because "mankind will have the enjoyable task of transforming the earth into a paradise." Man's "loving dominion of animals" (sic) will be a feature, and "the wilderness and waterless plain will exult." I am struck by the presence, in one of the pretty illustrations of paradise, of a bulldozer. Not my idea of heaven!

Nor is my purpose to make fun. The picture painted is a pretty one, and touching, in a way. But were such beliefs to gain many adherents, I would tremble for the stewardship of the earth.

What incentive could there possibly be to maintain biological diversity if you didn't believe in its mortality? In the same way, millions trembled on hearing Ronald Reagan speak of biblical Armageddon during the 1984 campaign. If it is inevitable, as foretold, what incentive exists to keep the finger off the button?

Probabilities speak louder than prophecies, but they both speak of annihilation if we carry on the way we have been. I would rather it didn't happen and support peace (and nonnuclear) activism for that reason. I see no inevitabilities as regards human behavior. However, should annihilation occur, I console myself that nature will persist.

I call this attitude a cosmic optimism. It simply suggests that nature, in the broadest sense, will carry on, having batted last in its minor-league game with us. We played catch for a brief while, dropped the ball, and threw a tantrum; whereupon nature took her big blue ball and went home to repair the scratches and scuffs we'd inflicted in its soft hide. We lost by default, and there were no more games in the season, for our season was finished. It mattered very much to us, but the rest of nature just didn't care, was rather tired of our company, thought perhaps we'd been a bad recruit to the league of species in the first place, and that she might not try that same experiment again.

That's supposed to make one optimistic? Let me put it another way, dropping the tired metaphor of a ball game like a high fly with the sun in my eyes (which was my first and last act in Little League). Imagine the sun in your eyes—your lizard eyes—through no smog. Imagine the lakes in your fish gills, fresh, pH 7. Conceive the cosmos untroubled by that spot of bother on earth, as all its peaceful, dumb species go back to their business of life and death and evolution, unperturbed by busy-busy people. I like these thoughts.

It would be dishonest to say that I feel no sadness at the prospect of the passing of humanity. Untellable sadness greets the very thought of it. When I consider the moldering of the last lost manuscript of Mozart; the combustion of the libraries when Fahrenheit 451 is reached early in the firestorms; the tumbling of towers and the crumbling of cottages, I could swoon (if I knew how) with earnest, dolorous regret. But think: all of the sadness in the world belongs to us. When we're gone, there will be no sadness, for it is a human conceit. So it would not matter, afterward.

The fact is, nature doesn't care. Only we care. And if we care so much, perhaps we should look for a few good platitudes to guide our critical actions in these days. T-shirts are a fruitful source of sayings: "Extinction Is Forever" is a good one; "We All Live Downstream" is another; and "Share the Earth."

"Cosmic discipline," John Hay wrote, "will not allow too much ignorance of what it cherishes." It is that discipline, finally, that lies at the root of my so-called cosmic optimism (just as our mammoth ignorance of what it cherishes makes me dread what comes next). And the "real stature" Hay mentioned, that "comes from an attachment with the unknown," I take to mean a buckling-up of our seat belt for the universal ride. Attaching to the unknown can be acceptance of nature, a faith in the course of natural events, even if they entail our own eventual extinction.

Taking satisfaction from such ideas implies a nonanthropocentric viewpoint. Copernicus saw that we weren't in the middle; why can't we? The natural world does not revolve around us, it merely tolerates us for a spell. We are indulged, yet we continue to indulge our own earthly xenophobia. Biting the land that feeds us, behaving like bulls in nature's china shop, and casting clichés across the littered landscape, we run serious risks. I am told the Finns around here had a saying, "You shouldn't shit in your own house." Taking it literally, they built their johns outside long after others had brought them in. We not only foul our own nest, we do it in the living room.

I can't help but keep on quoting John Hay, who employs never a tired phase and whose phrases never tire: "We have been cutting ourselves off, and we are wise to be alarmed. We have not been meeting the earth, we have only been erasing its opportunities, missing its indefinite, healing associations. Suddenly there is a terrible need for a great cognizance of the unity and interdependence of the world, in the sense of both human and natural communities."

There is another need, among progressive people, to realize that humanism can only take us so far down the agenda of "what," to quote Lenin, "is to be done." Beyond that, species-ism takes over. Liberation doesn't mean a damn in the face of imminent extinction, but if we can arrange a rain check on infinity, liberation means everything.

To return to heaven briefly, the popular idea of deferring it till later does nothing for our sense of obligation to the earth. The naturalist knows that heaven is here on earth (for those whose lives are neither too meager nor too glutted, too shackled or too free, to notice). The traditional view holds that there's more, and better, where this came from. I prefer to think that this is all we're gonna get, and it is more than enough, if we take time to experience it and care for it.

Beyond heaven and humanism, an evolutionary view is necessary. Evolution, we find, will adjust in rate and degree to the kinds of stresses imposed upon organisms. Extinctions will occur under stress, but so will resistance evolve, tolerances develop, and tactics adjust. Organic evolution will go on. As the only show in town, it must. Whether it goes on without us is another matter entirely. Eventually, it will.

Leading conservation biologists believe that opportunities may have ended already for significant evolution among large mammals, under the management regimes, stresses, and rarity we impose. Otto Frankel and Michael Soulé, in *Conservation and Evolution*, argue that this may be the case and that we bear the responsibility of preserving evolutional potential for as many species as possible. Whether we give them a chance to get back on the world by getting off ourselves, or take them with us one way or the other, will soon be seen. We still have some limited powers to affect the outcome. What stands certain is that we shan't arrest evolution much more without arresting our own.

Whether we choose to remain is our concern, and ours alone. Nature doesn't care. We are but a drip of spittle on the whisker of a beast in a constellation we can't even see. Nature has a right to care, and a sagging sack of grievances against our tenancy, but she doesn't. Nature gets along. Which brings me around at last to Willapa.

In the ravaged land through which we have been rambling, rarities have been lost, common creatures rendered rare, and the productivity of a great forest diminished for ages. The big trees and the bears are nearly gone, and the humans, many of them, are following. But certain species are doing just fine. Natives like the salal and the coyote thrive in the logged-off land, finding opportunities for expansion that they never dreamed of in the old-growth forest. Aliens such as the gaudy foxglove and the dowdy opossum proliferate still more, covering the clearcuts and the roads with their magenta blossoms and gray hides, respectively. These organisms evolved under stress; they know adversity and eat it up.

The weeds do it even better. Farmers with their sprays and archaic weed boards with nefarious powers battle gamely the tansy ragwort, Canadian thistles, and Himalayan blackberry. They make inroads with their powerful poisons. But make no mistake: the weeds will win: nature bats last.

In a sense, all life in the ravaged land is a bunch of weeds—survivors, coping and adapting under adversity. That goes for tenacious families who find something else to do when the creameries go under and the timber companies pull out, as well as for abandoned cats foraging at the local dump, and for huckleberries that clothe the clearcuts as soon as anything can.

Whether or not the weedy, faithful humans choose to remain, nature will not be crowded out, even here. Tonight I watched a possum waddle across the yard and up the slope to the road; every night, it takes what it will from our compost. We look forward to the marsupial's visits and hope it never has a date with a Dodge. I appreciate possums in the same way I admire starlings and cabbage butterflies and reed canary grass—not as native species, but as tough, clever, evolutionarily and ecologically astute organisms—as survivors, against all we dish out.

Last night the coyotes called by the covered bridge: first one tight, metallic yip, then a tentative croon, followed by five minutes of falsetto chorus in many parts, all countertenor and tremolo. "We are here," they say; "we'll eat your apples, your voles, your cats, the afterbirth of your calves; we're here, we set your dogs to barking, we intend to multiply. We are here to stay." That's what they say. Then silence, as they go about their wise, tenacious hunt for whatever there is. No other animal is more systematically or aggressively persecuted across the West. The coyote: evolving, getting better all the time, under heavy pressure.

In his book *Giving Birth to Thunder Sleeping with His Daughter*, Barry Lopez recounts a wide array of North American Indian tales of coyote. Coyote as trickster and in many other incarnations emerges from the pages as from the ancient campfires. I suspect one of those storytellers originated the last aphorism I wish to use.

Anyway, whether ancient or modern, it is a good one, and after publication someone will write to say they said it first. Watch for a credit in the second edition, should I be so fortunate. The saying: "When the last man takes to his grave, there will be a coyote on hand to lift his leg over the marker." The image should be struck on a new coin, with Charles Darwin on the other side; not negotiable, but a good-luck coin to remind us of change and evolution, and of creatures that will be happy to adapt if we ourselves cannot.

The land has been hurt. Misuse is not to be excused, and its ill effects will long be felt. But nature will not be eliminated, even here. Rain, moss, and time apply their healing bandage, and the injured land at last recovers.

Nature is evergreen, after all.

# Wintergreen

I began this book by describing an early winter's day as viewed from within the warmth of the Longview Public Library. Just a year later, I found myself in the same chair whence that picture originated, entering winter once more. We seek symmetry and continuity in our lives. In finishing these essays where they began, I felt a pleasing sense of both.

Last year, shortly after I wrote my concept of the wintergreen rain world, a dramatic snowstorm occurred. It made me wonder whether I had exaggerated to myself and to my reader the mildness of the maritime Northwest winter where precipitation is supposed to show in good, honest rain instead of sneaking into a snowsuit. Surely that blizzard had been a fluke?

This time as we drove over K-M Mountain toward Longview and the library, early November snow made the road hard to hold and the clearcuts almost pretty. I wondered if I was to confront a "real" winter's day from the library chair—a snowy, blowing day out there beyond the pane? Instead, we left the snow back on the Columbia somewhere around Stella, came into rain, and rode it on into a damp, gray day of sunspots chasing saucepan clouds across the sky. In other words, a normal winter's day in Willapa.

On the sidewalks that cross the town green, dense, wet moss filled the cement cracks like mortar. Those green stripes ran out to join the verdure of the lawn, little faded from spring, revivified from dry summer. Rosettes of the cat's ear, a wild dandelion, sprouted from fissures in the ponderous tumors of maple trunks. If you regarded the crowns of the naked eastern elms and oaks of the green, or stands of alders, stripped of leaves, on the near hills that showed between buildings, you might have thought "winter" in its New England sense. But if you considered instead the prevailing shades, in firs and grass and moss-mortar, and if you were from around here, you'd still think "winter;" but with an entirely different season in mind.

Winter—wet, green, slowly growing, cool, fresh, rainy, dripping winter. Only in winter, I believe, can one feel the full force of the rain that makes Willapa wintergreen in every sense. This year again, following my library session, a snowstorm set in—and lasted for an unprecedented two weeks! Yet when it melted, the green came back, only slightly bleached by the cold, white burden it had borne. My faith in the Washington winter was restored; snow or no, the gentle influence of the sea wins out.

Late last winter, deep in February, I drove a long circuit of the near rivers and ridges in order to win a sense of the season. I began in the Gray's River Valley by the covered bridge. Right away I saw a hundred or more mew gulls moored up against a pond's banks, looking like so many Styrofoam cups washed up on a shore left lonely after the regatta. Moving in every winter like a cold air mass, the small, pale mews worm the water meadows when the flood-waters drop. Nearby, from a salmonberry brake beneath spruce and maple, in a corner of the valley, a winter wren piped through the wet, green veil. Its song has as many notes as needles in the dripping hemlocks. Wrensong lends a sweet quality that instantly negates any dour aspect the gray day may possess.

The next pond down the valley, beneath the old Durrah farmstead, had buffleheads. The fluffball ducks bobbed, dipped, then deserted the surface before popping up again like black-and-white corks. As in the lily pond upstream, females outnumbered the drakes by two or three or more to one. Are the males shot more often? Who could shoot a bufflehead anyway and disturb that perfect plumage with shot, stain the silky white parts red? Some do. No hunters out that day to scatter the dabblers and divers.

Over the low divide and into the next drainage, down to Deep River. Driving upstream I noted the different colors of the bottoms—olive of the feggy *Scirpus* swamps; pastures brown-spotted where heavily trodden and muddy; yellow swatches of the reed canary grass, tall and unwanted in summer, dying back and still unwanted in winter; and emerald where low and wet but not so much as to be ponded. Straw, swampy swales of grass and brambles. The flats that hard labor won looked as if they're returning to the primitive state. Their cattle looked forlorn, sunk in cold mud to their knees. But skunk cabbages crouched in the pasture, preparing to pounce and steal the fields quite soon.

Up into the clearcuts, and the young, solid stands of hemlock and Douglas-fir where they have taken hold. Every hemlock owns a weepy, bent-over crown. Short, flat needles glistened with apprehended drops; wispy limbs seemed to show the very weight of the rain. As if more used to it, the firs stood more supple and erect, but their green seemed even more bleached out by the potent cleaning fluid of the rain.

The road climbed. The banks ran red, not Colorado maroon but orangy, as if even the land can't be left out in the rain without rusting. These banks are flat-faced, loosely lithified, ready to roll to the angle of repose. Deer ferns leaped: dark green fertile fronds, paler infertile fronds, dead deer-colored fronds of both types. Ferns, mosses, and conifer seedlings ran into one another in pattern and size.

These rivers, too, run close together, only the narrow wrinkles of the hills keeping them apart. The streams eat away at the edges, while the loggers work on the tops and sides. This must be how peneplains form. When erosion finally links all these streams, the ridge and ravine topography of Willapa will vanish, perhaps in time for the crust to thrust from beneath, creating the next great mountain range. But I couldn't wait for the peneplain; I had only hours of light left and had to cross the ridge to reach the next river, Salmon Creek.

Here, maples and alders leaned across the stream, the former almost entirely clad in mosses and licorice ferns, the alders daubed with green and the white lichens that make people think they're birches. Ferns stuck out at angles like poles with fish on, out, then down. Above, tall firs inhaled the mist, alder catkins caught the rain. The river ran bluish-brown between banks bright with the rosettes of saxifrages and foliage of buttercups.

From the top of an alder hung a six-foot sash of *Usnea*, a damper to mute any sounds trying to get through from the canopy. The tinkle of a golden-crowned kinglet, for one, tried to break through, but it was too high to catch. River's soft rush, rain's hard pelt. A great spruce was itself a rain forest of club mosses, each of many dead limbs a carpet show-room in the round. Fringes many inches long grazed the ceilings of the branches below. The mossy butts of other spruces made launchpads for tall sword ferns.

Small waterfalls sprung everywhere. With leaves off the trees, I could look across Salmon Creek and see the upper half of a forty-foot fall that would not be visible in summer: an upright freshet. The whole of the canyon seemed a great sponge, squeezed by the chain-mail glove of the glowering clouds.

Not many birds about. A ruffed grouse walked brazenly onto the road verge and moved in mechanical twitches as I approached and rolled down the window. Its rich nutty plumage spanned an inner spectrum of browns that the rainbow doesn't show. Then the bird finally saw me (the grouse would have been on its way to a potpie had I been a hunter) and it boomed off into the forest. Those drummer's wings can make an incredible noise when the heavy bird takes to its cannonball flight. (I once came home from a week away, to find a grouse dead on the living room floor, and a large windowpane shattered as by a stone.)

A varied thrush appeared on a limb, the bright sprite of the dunwood. This one brought to mind a favorite picture by Morris Graves. The Northwest artist painted many apocryphal birds and hedgerow creatures, but his varied thrush—rendered with just a few strokes in a field of olive shadow or vegetation—is instantly recognizable. Along with the wren and the kinglet, the varied thrush signifies the rain forest. The rusty smear it makes against the green as it flies and its nasal whistle leave you in little doubt that you're in Northwest forest, if you happened to forget. As I watched that day, I felt the only depiction of the varied thrush superior to Graves's was the very one before me then.

I parked beside a fabulous locust tree next to a spreadout spruce in a farmyard. Nearly every surface of the locust, all but the new growth, was clothed with mosses, lichens, and ferns. Licorice ferns sprang from the deep moss pillows, like palms through a jungle canopy viewed from the air. At least half a dozen species of mosses adhered to this tree. Yet the nearby spruce dangled only a wispy olive drapery from its needled branches, a kind that seemed to be absent from the locust.

While some mosses and lichens take root on any available surface (I make it a point not to sit too long in the woods) others choose a single sort of substrate, it seems. How do they select it, locate it, maintain it? And what keeps them from spreading to the tree next door? Microhabitat is a wonderful thing. We are far from understanding the whys behind the patchy distribution of many animals and plants. A tree such as this, in its patchwork coat of mosses, shows just what I mean: who picks the pattern?

Natural selection, is who. Epiphytes arise, suited to each sort of surface in the rain world. Some specialize and do very well under limits; others generalize and do fairly well all over. Tolerance for light or shade, wet or wetter, have much to do with the ultimate arrangements. Those olive wisps on the spruces strike me as shade-loving; behind, in full sun, limegreen lichens decorated an alder in shabby style, hanging all over in tatty scraps and sashes. There are partial reasons that could be measured and modeled. But beyond that, the sociology of plants in close quarters seems to evade analysis. God's landscape plans have been lost. Confusion reigns in the garden patch of Pan. Amen.

We lack, at least, the bromeliads of the tropics, those phantasmagorical air plants that crowd every inch of branch like the Professor's burgeoning desk in the comic strip *Shoe*. The closest we come is our *Usnea longissima*, whose long, flowing tresses somewhat resemble the common bromeliad of the American South, Spanish moss.

Between this locust and spruce (their upper branches intermingled), and their neighbors, and the mossy-topped cedar fence posts between their trunks, we have a powerful pastiche of Willapa: native and alien, shades of the pioneer, plant on plant, the perpetual green products of rain and rot and regrowth.

I continued down Salmon Creek. Beyond the locust, a field gone to wild cherries showed bits of lichen in mahogany wet branches above rusty bunches of bracken bound with bramble twine. Dull but richly earthy colors. More: Oregon juncoes, alarmed by my arrival on the scene, flicked across the road, call notes ringing, white tail feathers flashing. A cheeky one perched and looked back. Coal head, russet back, beige sides, pale pink bill—just a junco but a prominent spot on the moment's record. One feels sometimes it's all green, it's all too green; but the other colors assert themselves, if modestly. Winter hues, relief from green.

The sun slunk low; the day grew duller. I crossed the next ridge, over the Crown Main Line, from Salmon Creek to the Naselle River. These roads, along with the stumps, are among the few palpable benefits left over from logging's extravagant past. People like me use them frequently and see them as token repayment for the damages inflicted on the whole Coast Range. We've no roadless wilderness left here, so we may as well explore through the agency of the roads that took it away. This route, like many in the area, used to be a logging railroad. No sign of trestle or track remains, and an old tunnel on this grade has been blasted shut. It is as if, having created a historical resource in place of the natural, the companies couldn't wait to remove that too. The rain and the rot were all too pleased to fall into league and erase the last traces.

I call this crossing the Lotus Road, for it has the only big stands I know hereabouts of *Lotus crassifolia*, the thick-leaved lotus. As green as the rest of the itinerary, the Lotus Road may be greener yet in May; this legume is the host plant of the bramble green hairstreak butterfly, a bright emerald mite that I have yet to discover in Willapa. This lotus-eater and its host are two bits of verdure one cannot find anywhere in winter, dormant as they are in chrysalis and seed. Beyond the bare bank where the lotus grows in May, out of salal erupts a cascade of yellow club moss, like a leak of molten gold flowing down onto the road.

As if stricken by the eruption, the drawn bodies of horsetails lie supine against the pebble bank. Having been green all spring, summer, and fall, in the green winter the horsetails bend and fall down the banks on their faces, finally lying dead, dried, and bleached, with inky black zigzag nodes like charts of vital signs run down. They remind me of so many many-jointed bones, stretched out with long tufts of sodden brown hair still attached. Hard to imagine that from these will spring some of April's most urgent greens.

Below me, then, the Naselle River. Across it, the panorama of the hills, looking looming, as if struggling to break out of the mists. Interwoven patches of conifer greens and alder browns, fogs drifting in and out of every fold and hollow and blending them together, breaking down that angiosperm-gymnosperm barrier with endlessly intertraded seeds of rain. Seventy million years of evolution or more, blurred in a single day's downpour. Alders in the foreground hung heavily with new catkins, old cones, and ageless raindrops.

I came down to the Upper Naselle, where the road crosses a high bridge and breaks into three to head toward the North Fork, across the hills on the Bonneville Line to the Chehalis, or down the Nemah to Willapa Bay. The opposite way leads down to Naselle, and that's the way

I went. Across the bridge, tall, slender alders over the gorge bank had arms all musclebound with moss and long, blond locks of lichen hanging nearly to the black water. In summer, when the river drops to expose the pebble beach below, neither the moss nor the lichen may be fully appreciated. Then they lack something of their winter's luxuriance and hide behind the season's new leaves.

Now, in winter, the fallen leaves signal a rest from growth, a dominance of death. In Willapa the winter wetness equals life and growth in partnership with death and rot. The volcanic reaction called photosynthesis slows down but never stops, as plants squeeze scant sun from the dull, short days. Even the great logs that form the bridge's safety sides have grown so waterlogged that minute hemlocks, Christmas trees for beetles, spring from cracks in their slick brown sides. The real secret of wintergreen is the sea's gift to the sodden land: mild temperatures and abundant water. Winter gardening is popular in the maritime Northwest, but it's not really gardening: it's hydroponics.

One sign of winter, less welcome than the monsoon, just as true, greeted me as I descended the Naselle: the sharp increase of litter in the woods following the incursions of the hunters and steelhead-salmon fishermen. That someone who supposedly loves the outdoors can trash a sylvan spot simply begs belief. The Model Litter Control Act, which the beverage bosses keep rolling out like a barrel to crush the bottle bills that arise from time to time, is a crock; and in any case the occasional clean-up crews it pays for never reach the county roads, let alone such woodland lanes as this!

Many hunters, one observes, find it impossible to keep a can in their possession; the act of throwing it out must be an expression of manhood, killing large animals with high-powered rifles not sufficing. They invariably throw every third or fourth shiny can into the salmonberries, where it offends but cannot easily be retrieved.

More offensive yet on a carpet of moss, to be sure, so visible, so violating. I filled the Honda's hatchback with aluminum, but the large wine jugs I recycled into the river; they may reach the sea, or sink to the bottom as a hideout for fry.

Again I visited a favorite miniature rain forest above the river. A few old-growth spruces have somehow survived here, perhaps because the owners know that this stretch of river with water-carved, blue-green rock pools, known as Cut Rock, is a favorite spot for locals, for fishing, swimming, picnicking. An outcry would surely follow logging here. But that's the price you pay for living in a loggers' land: you grovel for the scraps they toss and learn to live without the rest.

Flowing water pounded, muted; dripping water pinged on slick leaves and fern fronds. Alders lay a white drop cloth for the greenwashing of the world.

The light diminished, and with it the intensities of green; but the shades remained many in their soft subtlety. Moss-hung arms and stumps stuck straight out from spruces a'midtrunks, dripping pitch. Looking up, I saw the spruce canopy, almost as gauzy as the moss. Winter permits conifers, as well as epiphytes, to come into their own. That certain mosses resemble the graceful boughs of western red cedar you might not notice in June, when the leaves of the hardwoods and herbs encurtain both cedar and moss. In February, the competition relaxes; flashy plants withdraw to rosette or tuber, leaving evergreens in charge.

A star-carpet of floor mosses like Lilliputian tree-ferns held two little hemlock seedlings: one was sparse with needles, heavy with mosswebs. The other grew free, was bright green and lush, yet they grew side by side. Why? Appropriation of dappled sunlight, perhaps; a root-nibble by a tunneling rodent? The limits to growth brook no backtalk from plants, even if humans seem to ignore them.

Sword ferns show no awareness of limits, shooting slick and riotous from the moss mat. They burst out from their stools into five-foot, gamboge-green fireworks imported from green-seeing Oz. Sword fern: supplier in season of spores by the bagful, merchant in chlorophyll all the year round. Sword fern, trademark of the rain forest. Rain forest, wintergreen's apogee.

Once out of the big spruces I came to one of my favorite stumpfields. A tall hemlock nursed from a high cedar stump had nearly eaten it with its stout aerial roots, tropical-fig-like. Many things seem tropical here, even the clearcuts, which resemble the devastated tropical rain forest prior to the planting of oil palm plantations. But there was nothing tropical about the temperature, in the high forties or low fifties from which it might not vary in twenty-four hours. I followed the Naselle River down into the valley, leaving the rain forest behind; but not the rain.

The rain had fallen nearly all day. Then, before it set, the low, post-equinoctial sun sneaked out of the cloud bank and dissolved the rain. Blue spread. Sun struck wet ponies, made them shine and strut. Crows spattered a pasture like black fen turves, punctuation marks for winter's concision.

I hit the highway and turned east for home. Slick osier dogwoods in the roadside ditch glimmered like ruby wands. All the colors intensified in the last stray bolts of sun. Steam rose, wraiths wandered the valleys in search of damp retreats. Spring seemed imminent. Then the clouds began to close again around the sun, which set in any case; rain resumed, colors faded, and the green, wet winter went on. Night fell on the February fields.

The river route showed me again that rain makes the world go green. But rain also sponsors rot. Without rot, no renewal; without rain, neither. So as the winter drips through, the great compost heap of the land cooks away, preparing its own spring feed. Whether the harvest of the leaves comes all in the fall as in the North, all year around as in the tropics, or some of each as here, there can be no new buds without the silage of the old.

The nature of rot and renewal are little known to us, and we like to keep it that way. Rot, we do not wish to accept into polite company; only growth seems an acceptable state. This is another denial of death (we have so many) but more than that it is a denial of life itself, for without decay there can be no regeneration. No room, no food: we organisms die to make room for the next and to feed them on our leavings.

People used to be allowed to take part in the process, only a pine box between them and the soil. Modern bodies become sealed in cement cases, hermetically, following chemical preparation of the body for the fiberglass coffin. In this way we deny our elements return to the earth until orogeny itself breaks and spills our concrete cocoons. This is my idea of perfect purgatory. So many die, yet so few are allowed to feed the future. Northwest Indians arrayed the dead in tree platforms, to speed the process; we ingeniously delay it, as if to frustrate forever the human compost. Why the terror of the worms? That way, through the gut of worms, say I, lies heaven: to be recycled in a worm casting, to feed the green grass itself.

In the same way, when the great logs are taken from the forest, the forest is denied its refreshment. How can any organic system go on, whose bodies are robbed one by one? The soil cannot put out forever in the absence of composting. Artificial fertilizing of forests isn't the answer; it just means robbing from one part of the corpus to feed another. The forests need fertilizing by their own kind. Something more needs to be left behind than commonly is, or the soils will stop putting out.

When I empty our chamberpot and eggshells into the compost pile, it feels like a small rite. To keep the atoms in flow, that's the idea. It is as much to keep the aluminum flowing that I fetch it from the littered woods, as for looks or pennies per pound. Recycling is one of the few completely good things there is, as much for the pure idea of it as the resource sense it makes. And when it comes time to recycle ourselves, we should revel in it. Morbidity comes from holding on to bodies; letting go is not morbid at all, but cause for celebration.

Perhaps our inability to understand the need for renewal comes from our visceral rejection of its flip side, rot. Some go to the desert to desert rot altogether, only to end up mummified. We who are left behind in Willapa cannot avoid it. Rot's agent is named damp, and the damp demands attention. In England, another wintergreen land, the natives complain of "rising damp" in their walls and employ contractors to install "damp courses" to prevent it. Perhaps, with gravity on their side, they can succeed in keeping down the rising damp. But short of a futuristic dome, I cannot conceive of the damp course that would hold back our dropping damp.

When decay overcomes renewal in the cycle of the seasons or in the course of lives, a state of senescence sets in. This happens every autumn in Willapa, for although green prevails, active growth slows down and the rate of rot picks up beneath the dead leaf layer. So the well-trodden metaphor of autumn for aging makes sense. Senescence, in fact, means growing old. To senesce is to age, to slow down the vital processes, to prepare for death. But senescence has another meaning in biology, since a system never dies, it just changes. It means to age, all right, but to be reborn in the process. A lake senesces by filling in with organic matter and ultimately becomes born as a bog, then again as a meadow, then as a forest. Senescence holds terror or sadness only when it's a dead-end street. And there are no dead ends in nature, since mortal individuals = species = matter = energy, which is immortal. Senescence, then, is relative.

I like a quotation in support of the definition in the *Oxford English Dictionary*: Woodward, in his 1695 *Natural History of the Earth*, wrote that "the Earth, Sea, and all natural things, will continue in the state wherein they now are, without the least senescence or decay." Of course, the hoary monographer had it exactly wrong, since all natural things will and must decay and regenerate. But I believe that what he had in mind was the continuing vitality of the whole, which, as long as the cosmic climate permits, is exactly what all that decay and growth balance out to preserve. When our sun becomes senescent in absolute terms, an eventual condition that Woodward could not have anticipated, it will be a whole new ball game in which you-know-who bats last.

A climax forest exists too in a senescent state, where aging and rot roughly balance, or may even overtake for a time, regeneration. I suppose it is the mistaken idea of very old-growth forest aging toward ultimate rot that gave rise to that sillyism some foresters spout about "overmature" forests. We know, of course, that climax forests age like good brandy—they just get better. The wonderful cedar grove on Long Island in Willapa Bay demonstrates this clearly: four thousand years without significant disturbance and getting stronger all the time in its old age. In the fullness of time, natural factors open glades and thus prospects for successors. Forests did manage to replace themselves, somehow, prior to the advent of modern forestry.

You have to look hard to find anything approaching a climax forest in Willapa, the Long Island cedar grove being perhaps the only example. But a different sort of senescence is much more apparent, everywhere you look around here in fact. That is the senescence of a culture.

Societies tend to remain emergent for centuries, viable for more centuries, before becoming decadent and finally being replaced by other societies: much like forests, which are societies also. Along the way, they attain periods of maturity that may be lengthy and rewarding, after the disruptions of early growth and prior to the throes of demise. This is simplistic, but we have some cultural models for such a pattern. They suggest that protracted senescence can be one of the best periods in the life of a culture.

There was no time for such a genteel evolution here in Willapa. As if afflicted by some disease that strikes down youths in their prime, southwest Washington culture barely reached adolescence before senescence set in. Now, apparently accelerated toward the latter stages, we watch for a renewal that may or may not come.

I could be on thin ice here, taking liberties by drawing parallels between cultural and biological situations, and telling my neighbors, sons and daughters of the pioneers, that their culture is old before its time. Rot, decay, senesce: these are strong words. And as I've already suggested, they are not popular concepts in our modern, youth-dominated, growth-oriented society. No one can deny that we have economic problems in southwest Washington; but senescence? Isn't that a bit extreme?

I don't think so. Just look at the evidence. Houses and barns rotting into the brambles at every turn in the river, and many of those that aren't abandoned, for sale with no takers. Businesses shut. The Cathlamet Hotel, finest structure in Wahkiakum County next to the courthouse, standing empty and in decay. (We attended the auction when its furnishings and fixtures were sold: it was like witnessing a public disembowelment from within.) The woods abandoned for now by "the tree-growing people," their former faithful out of work on all sides. It's not a pretty picture. Today's paper carried a letter about a family, hungry, about to be had up for possessing a poached elk. There is poverty in the High-Yield Forest.

I think the most striking demonstration of senescence comes from a comparison of the present with the pretenses of the settlers. Old photos in Carlton Appelo's telephone books or in the museums show men in suits and women in broad hats, bustles, and all manner of lacy finery, posing in front of huge, whitewashed new stores and hotels, on the decks of the daily packet steamers from Astoria or Portland, perched on immense stumps of old-growth firs, or, stiffly, standing on the porches of New England-steepled white churches.

Little if any of this may be found today. Their descendants may be seen instead wearing nylon windbreakers with tavern ads on the back, chatting colloquially by their pickups at the gas pumps. Nothing wrong with that, styles change. But the trappings of genteel society preserved precariously on the forests' edge bespoke something more than late-Victorian fashion. Along with the many theaters, newspapers, schools, churches, lodges, civic and cultural organizations, they spoke of hope. High expectations. An intent to create the culture they knew in Stockholm, Helsinki, or Chicago here in the Great North Woods. It's not that nothing of this remains, but that it has so contracted as to give one an overwhelming feeling of its being well past bloom.

I have before me a 1979 reprint of a circa 1910 booklet entitled *Pacific County and Its Resources*, a small masterpiece of boosterism aimed at luring investment and settlers and raising home pride. The rise in the booklet's price from $.25 to $4.50 says something, but a comparison of the period advertisements with the current yellow pages tells much more. The sheer number and variety of extinct businesses in the northern and western parts of Willapa are astounding. Accommodation and transport alone make it sound a better time to be alive and exploring the area, which was doubtless more interesting then in almost every way. And there were hotels, guest houses, and restaurants in every community, regular public transportation by steamers and trains over much of the region. No doubt life was much less convenient then and imposed many more hardships; but it also appears to have been more colorful. Trains and steamers—it is difficult, and quite painful in contrast with our automotive monotony, to picture!

Advertisers proffered their goods with a sense of beneficent pride seldom heard today. C. B. Handy, "Universal Providers of everything to eat and wear," resided in Lebam, now but a pleasant wrinkle in the highway east of Raymond. Next door in the Willapa Valley, tiny Frances boasted the Custer Mercantile, dealers in "Fancy and Staple Groceries, Shoes, Hardware, Grain and Feed and Agents for the Celebrated De Laval Cream Separators and Dairy Supplies and New Royal Sewing Machines." And these were the small settlements—South Bend, Raymond, Naselle, and the Wahkiakum County towns offered many of the goods and services we drive one hundred or two hundred miles for today.

It would be unkind to continue making invidious comparisons based on an ingenuous spirit of enterprise, civic optimism, and faith in the future of the country and the region. Unkinder yet to quote the descriptions of each of the young towns in the county. Just one heading will give an idea of the pathetic optimism that beat in every heart in the Willapa of 1910: "South Bend—The Baltimore of the Pacific."

The founders had good reason to be optimistic about growth. The seemingly unlimited supplies of timber, water, power, oysters, fish, farmland, and other resources promised no end of opportunity for expansion. The booklet tells us that, in 1910, Raymond had a payroll larger in proportion to its population than any other city on the Pacific Coast. How could they have foreseen the mill closures and chronic, acute unemployment that lay in wait three-quarters of a century later? Opera houses and dance halls situated in remote logging towns displayed the people's unwillingness to be without the amenities they considered proper to civilization, as well as their belief that true civilization would surely arrive if only they made it so.

But the big timber ran out. The native oysters were depleted and the transplants proved problematical. The rain discouraged settlers who were supposed to be enamored of the region's "antiblizzard, anticyclonic" conditions, lacking a "single death from any malarial disease," and possessing "varied industries." At that time the industries were indeed varied. Farms proved productive in places, but the lack of good transportation to the cities and other factors caused market centralization, the closing of the area's excellent creameries, and a sharp reduction in the number of dairies. Many of the anticipated crops never materialized in market quantities, cranberries proving the strong exception. The sublime, benign region proved less tractable than had been thought.

What went wrong was less the fault of the rain than that of the overeager takers. With "miles of deep water frontage, free mill sites," as one panoramic photograph advertised, timber giants moved in, co-opted the labor force, undermined diversity, and left the towns high and dry when the logs and jobs ran out. The author of the booklet can be forgiven his exuberance in the spirit of super-boosterism that marked the times. But he simply misjudged when he wrote, "The year has never come and never will come when the lumber, fish, oyster, dairying, general farming and summer resort business . . . are all in the dumps." He believed, of salmon, timber, and the rest, that the "supply can be maintained provided the advancing wave of civilization does not engulf the [industries] with its destructive tendencies." That "advancing wave of civilization" never came, but the "destructive tendencies" evidenced themselves just the same, and now it's the dumps after all.

Willapa had already just about peaked when that booklet was written. Of course, there were further developments and quite a prosperous period following World War II. But the anticipated metropolis never came to be, and the trend has been steadily downward for many years now. Rather than the prospectus for future growth that it was intended to be, the book became a wistful chronicle of a fleeting period when growth had gone about as far as it was going to go. Hence, its recent reprinting.

The signs of senescence are all around. A more conservative expenditure of the resource base might have permitted a longer wave in the cycle of growth, and a few more of the dreams might have come true. If not the new Baltimore, then at least a stable city of unpretentious dimension. Even if the second-growth timber had been fostered in the fifties, instead of being cropped still more intensively than before, the forest-products industry might still be on its feet in Willapa. But market factors beyond the subregion have played a role, and these might have thwarted the growth of a new Bay Area just the same. In any case, the boom burst, and now senescence reigns. To all appearances, Willapa is going back to nature rapidly.

Now let me backtrack a moment. How bad is senescence? Would anyone who lives here now really like to see another San Francisco filling Willapa Bay and spilling over the hills? What's left is not so bad. We have the benefit of still better services half a day away by auto, though it would be more fun to be able to take a packet steamer still. One or two spired churches have been restored; new ones have arisen. Schools have been consolidated and might be better for it. The minute marts lack the charm of the mercantiles, not to mention the selection, and the general stores that remain have long since given up the range of "fancy goods" their predecessors carried: no more cheese wheels or pickle barrels. But we lack nothing necessary to life, and we have the bonus of an interesting (if brief) history bequeathed by the big dreamers.

Life in Willapa is generally adequate; for some, more than that. We have pulled in our horns, diminished our expectations, and made do. In a sense, that has prepared us for the mood and realities of diminished expectations that have overtaken the country as a whole in recent years.

Population, that perennial measure of prosperity, no longer works as such (consider Mexico City). So when we confront the depopulation of this region, we needn't consider it a blatant failure. In fact, failure needn't enter into the discussion, beyond a recognition of failure to steward our resources. Yet the results translate into failed businesses, lost farms and homes, and emigration. Jobs go, people go, and don't return. As Bruce Springsteen wrote in "My Hometown," "Foreman says/these jobs are goin', boys/and they ain't comin' back." In fact, the whole area has something of a Springsteenesque feel to it these days—mills closing, the young leaving, a melancholy refrain.

Yet, there is something else here that is distinct from all that. When the mills close, the log trucks stop rolling, the dairies go to summer beef or weeds, the fresh, wet green remains the same. It is something the mill towns of the East may not possess. The essential beauty of Willapa has been marred, but not erased. The quiet can only enhance the appeal of the landscape. It keeps stubborn people here and brings new ones. While depopulation inevitably followed the successive turndowns of industry and families have been hurt and all but exiled, others, better suited perhaps to the new conditions, will remain. That's evolution.

One of the difficult things for people to realize is that senescence (a prettier word for rot, reminding one of the susurrence of the wind) is not all bad. In ecological terms, senescence must precede (and is normally accompanied by) regeneration. All communities must senesce in order to be renewed. Renewal can occur continually in a climax state, as in the Long Island cedar grove or in New York City; or it can take place episodically, as following a forest fire or a bad recession.

A mature rural economy, like that of England prior to World War II, is one in which senescence and regrowth occur together: a kind of a climax state of the countryside and its settlements. Willapa never had a chance to get there, or to come even close. Instead, it followed another classic pattern in ecology: boom and crash. It is an uncomfortable way to go, it hurts people and their pride. But it leaves open the door for stability next time through a different model from the one that created the crash.

So we have senesced without the concomitant regrowth that denotes gentle maturity. That way lies obliteration. If we are not to fade away, we need to begin to grow again. But we would be wise, I feel, to avoid setting ourselves up for another crash—not that there is much worry of that. The chamber of commerce types have a hard time buying the "senescence is okay" argument, as do the displaced families. They may cling to the hollow husk of hope that another boom will come. But theirs is a faith in what's not going to happen. They are the casualties of simplification.

We know that a diverse ecosystem is a stable one. The simplification of the Northwest woods will lead to instability, just as lack of diversity in the workplace led to the unstable present situation. Those early boosters had it right when they plumped for a highly varied agricultural and industrial base. Had it worked out that way, their plan might have come to something.

Now, in our winter of disconsolance, a new diversity is beginning to arise. Drawn by the very features that represent economic decay—empty old houses, lower land prices, the uncluttered countryside of the largely abandoned valleys—a new sprinkling of settlers has been arriving in Willapa. They include the naturalists I wrote of, as well as retirees from the merchant marine or the military, industry or the office; yuppie dropouts and seasoned hippies, all anxious for a taste of the land; young farmers and traders and technicians willing to work hard to buck the tide; craftsmen and -women, artists and artisans. Natives, having left for the lure of the city or the lack of a job, come back to make one.

That is the essence of the new colonists: they make their own jobs with an astonishing range of skills and resources. In so doing, they begin to bring back something of the living human diversity that's been lost.

Meanwhile, the old-timers who remain, the Finns and Swedes and Swiss and others, furnish the essential matrix for regeneration. That's what an enormous clearcut lacks: old-timers to tell the tales, to pass on the locally adapted genes and knowledge and skills and traditions that successful regrowth requires. The purge of the local population began to look like a clearcut, until it became clear that a good many of the locals were loath to leave. Some, having become prosperous off the land when it was still giving, have kept their money here. Others remain on scant resources, bolstered by the nearness of family and friends. Family looms large in Willapa society. While extensive intermarriage can prop up provinciality, it also lends texture to the community. As newcomers arrive, the incestuous features of the towns and country give way to their refreshing influences; still, the strength of families remains.

Among the rotting homesteads one sees many tidy homes in good repair, proof against the agents of senescence, ready to shelter another generation from rot's best friend, the rain. Nothing shows the evergreen possibilities of any region like those who stay, for they are the ones who have seen the boom, lived through the bust, and still believe in the land.

One still sees mobile homes plopped down beside fine old houses left to return to the humus. It's cheaper, and a lot easier, than fixing up the old. But just a few of the original homesteads are being reclaimed, lovingly restored, or remodeled into places where life can take over again. The germ cells of regeneration have been planted; now it's up to initiative and evolution to decide what happens next.

Fortunately, the infrastructure is still intact. Wahkiakum and Pacific counties, the latter bolstered by the still fairly prosperous Long Beach that lies outside this book's concern, have been well enough managed so that essential services remain intact. Water and electrical systems, roads, schools, and such still work. Pacific County has a bus system, and

Wahkiakum has a ferry from Puget Island to Oregon, the last ferryboat on the Lower Columbia River. This is no string of ghost towns along U.S. 101 and the Ocean Beach Highway, yet. But they're running scared. Everyone sees the signs of senescence, and wonders how to get young again.

A comprehensive plan for Wahkiakum County was adopted over great resistance. People saw it as a barrier rather than a tool for intelligent regrowth. They did not realize that without a plan you are forced to just sit back and let things happen to you. The plan was drafted at the state's insistence, but it is so general that it cannot accomplish much on its own. At least it states the majority of residents' desire to keep the county largely as it is, while creating new opportunities. A citizen advisory vote to reject a nuclear power plant some years ago expressed the same attitude. Few of us wish to exchange our depressed condition for prosperity at the expense of that which makes it worth living here: Bhopal instead of Appalachia is no deal.

In line with a similar movement statewide, the Lower Columbia Economic Development Council arose to promote appropriate growth. Efforts are under way to lure light, clean industries to the area. Many people feel that a more realistic tack would be further development of the tourist resources, hand in hand with cottage industries that would appeal to visitors. A survey was taken of services people wish to see here and which they would patronize. Suggestions for new directions have ranged from Finnish food factories to a real-ale brewery and an arts center, utilizing derelict structures such as solid old creameries and immense barns now standing empty.

Little of this has come about yet, but bed-and-breakfast houses have been opening, and a crafts center, a gallery for local artisans, is planned for the historic Redman Hall, an imposing structure in Skamokawa's National Historic District. The small but strikingly situated Cathlamet dock is to be restored in the suitable milieu of a Norwegian waterfront in hopes of emulating the highly successful ersatz Bavarian village of Leavenworth in Washington's Cascades. The Columbia River flows by, waiting.

Things are happening. Congressman Bonker came to the county to open a computer-cover factory. The business, which moved to an old farm in Deep River from California, will create twenty jobs. Silicon Valley it is not—but it's something. Even as families continue to leave and businesses close, new ones come and open. That is the nature of senescence: death and replacement, rot and regrowth. Those who want to be here badly enough will find a way to be here, as friends of mine laid off from Crown and Weyerhaeuser have told me. They have faith in their skills and willingness to adapt. They will find an increasingly diverse community evolving around them, without the stresses and impurities that realization of the old dreams might have brought: our air and water are still very pure.

In a senescent state of affairs, specialists survive. Climax forests support specialized organisms that do very well there but could not compete in the rat race of regrowth. That's why species such as the spotted owl become endangered when the old-growth forests disappear. As a community matures in late senescence, so it is with its occupants. Those who succeed will be those who can do something very well. A few generalists, jacks-of-all-trades, will make it around the edges. In a period of rapid regrowth, the reverse takes place. Adventive, opportunistic types thrive, while the specialists must move on or face intense competition that they may not be able to withstand.

Earlier I said that the people here were weedy types, able to adapt quickly to available opportunities in a disturbed situation. But as senescence advances toward a new maturity, the survivors will tend to specialize. While retaining a degree of elasticity, they will refine their functions to fit closely the niches they have seized. It has to do with putting all one's efforts into skills for fitness, as opposed to skills for rapid expansion. Up in the logged-off hills, generalist organisms are doing their best to rebuild the shattered structure of the forest. If they have time to succeed before the next shave, specialists will return and contribute diversity. Down here in the senescent valleys, the new specialists are beginning to make it, just. This may not be a bad time coming up for them.

People speak of economic recovery everywhere these days, as if it were the new grail. It's the same in Willapa as in Springsteen's New Jersey, and everywhere else the boom-bust cycle has ruled. Boom-and-crash may suit the shifting needs of capitalists and locusts, but most species and most people are happier in a more or less steady state. Sufficiency, security, and modest expectations for growth: these are to be desired. Change and trauma enough will occur to provide renewing influences. Peaceful, creative stability makes more sense than rapid growth that outstrips its resources and is bound to bust again and again. In this sense, Willapa makes a model for any crashed community in the postindustrial wasteland.

If the planners succeed in attracting more opportunities for those who are already here and encouraging the settlement of a variety of newcomers with skills and services and talents to offer, then little will change except for a subtle strengthening of the fall-back position we find ourselves in now. A more interesting, more stable, and more mature community would slowly grow.

If they do not succeed at all, nature will slowly reclaim the region. The towns will grow still more skeletal and the countryside really depopulated, run by a few absentee lairds, containing mostly beef cattle to be shipped in and out according to the growing of the grass; a sparseness of farmers; a seasonal flush of hunters and fisherman; and a low level of logging. Any culture that remains, such as it is, will surely die out.

But if economic development should succeed in stimulating a dramatic recovery, through the advent of major new industry, so that booster time is here again and crazy high hopes spring internal, then everyone involved should know what changes would be in store. They would mean more than jobs, taxes, kids for the schools, and customers for the shops. In a period of active renascence, ecologically or socially, great disruption occurs. The adventive species move in and frequently displace those that were keenly adapted to the gentle state of senescence. Just look at Rock Springs, Wyoming, or any of the new energy towns of the West: they have gone from quiet and peaceful to busy and trouble-plagued. And some, like the oil shale new-towns of Colorado's western slope, have already crashed. Who needs that? Someone always profits in the boom-bust cycle, but it isn't usually the local people.

The likelihood of that happening here is so remote that I scarcely worry about radical change coming to Willapa. Somewhat more likely, and a far healthier prospect, would be modest regrowth based on appropriate use of the fallow land. The need is to reinforce the qualities of the land and its people with immigrants and visitors who come on account of what may be found here; then to keep it that way, and enjoy the results. This way we could someday regard our senescent state with happy heart and goodwill. Bishop William Stubbs in 1884 observed, about the English tongue, that "It is not a dead but a living language, senescent, perhaps, but in a green old age." Is this not a congenial way in which to regard a language, or a landscape?

What we want, then, is to develop a kind of permaculture. This term was coined in 1975 by Australian landscape ecologist Bill Mollison. It describes a state of "sustainable land use within the context of a sustainable and humane culture." Mollison believes this can be accomplished, on different scales, by "designing ecosystems that are food and energy producing while conserving of resources and wildlife habitat." If ever there was a place in need of permaculture, it is Willapa today.

Without calling it permaculture, Richard Mabey described such a state, once prevalent in the English rural landscape, in his book *In a Green Shade*: "There is a sense in which a settled rural landscape, whose pattern of fields, farms and churches embodies the history of a hundred generations, is a vision of Eden, no matter what temptation and toil lie behind it." Willapa, with only four or five generations under its belt and running down fast, has a long way to go toward such an ideal. We may never get there. But if we should, the rewards would be great.

Whether all else changes, stays the same, or just fades away, the seasons at least are immutable. Of all four, the green winter here makes the deepest impression, through the sheer persistence of its pervasive, dripping, rising and falling damp. The rivers flood across the valleys, and the buffle-heads and mew gulls settle in, as mists descend into shaggy hemlocks and pale green lichen seems to swaddle the world against all danger. Green drips into deeper green. This is winter in Willapa, where the pelting rain brings the promise of recovery to the bruised hills. Later, when the washing relents, the brighter shades of spring green always arrive.

Will the people be refreshed, along with the used-up year and the ravaged land? Maybe, maybe not. But at least the seasons will survive. And I suspect Willapa will as well.

# AFTERWORD

*Willapa Since Wintergreen*

I've come inside, reluctantly, to complete this revised coda for an old friend. My reluctance is not for the writing, or for the friend: I am delighted beyond words that *Wintergreen* is making a reappearance in this beautiful new edition, and nonetheless happy as pie to put down some new words for the occasion. My reluctance comes from the fact that five different species of butterflies are flitting about out there, nectaring on cherry laurel, Asian pear, salmonberry, and bluebells—in the middle of March! I've never seen such a thing here before. Willapa, like the World, is warming, and the land of wintergreen is not the same place it was when I wrote this book, thirty years ago.

Yes, thirty years have passed since *Wintergreen* first came out, fifteen since the last paperback edition, and I'm still here. No one is less surprised that things have changed in Willapa; perhaps no one is more surprised than I at the nature of some of those changes. Now, with the publication of this new Pharos edition, I have the rare opportunity to reflect on the seasoning of my previous thoughts, the metamorphosis of my subject, and the outcome and aftermath of some of the stories I was obliged to leave hanging in earlier editions. I have fixed some errors and infelicities in the main text, but have made no attempt to alter the original time frame.

Some details, therefore, will seem dated; but that's the way they were. Hence this afterword, to bring things up to date. Not that this is the last word. In real communities of humans and other species, sharp endings seldom occur; life is preparation for change, and the stories go on and on, like the land.

First, how has the book held up factually? I am happy to say that no serious challenges have arisen to the main facts in *Wintergreen*, although not all of them were welcome in some quarters. My accounting of "the sack of the woods" has withstood scrutiny at the level of what happened and what didn't. My opinions and conclusions based on those facts, of which personal essay must largely be made, differ of course from those of some readers. But on the whole, I feel confident that the reportorial side of *Wintergreen* can still be relied upon by careful students of Northwest land use, history, and politics of ecology. Several fine, more recent books, including William Dietrich's *The Final Forest*, Robert Heilman's *Overstory Zero*, Kathie Durbin's *Tree Huggers*, and Chris Maser's *Forest Primeval* all help to bring the saga of the deep dark woods up to date. James LeMonds's *Deadfall*, sprung from the opposite corner of the Willapa Hills over by Castle Rock, gives the best picture yet of logging life (and death) in southwest Washington.

When the first edition of *Wintergreen* was published (Scribner, 1986), we knew of little old-growth forest left in the logged-over landscape of Willapa. I highlighted the already-famous Long Island grove of ancient cedars and a small but rich remnant called Hendrickson Canyon. When we left the former, Congressman Don Bonker (D) was determined to wrest the money from Congress to save the second half of the cedar grove from cutting by Weyerhaeuser, who were holding out for top dollar. With Senator Slade Gorton's cooperation in the Senate, he succeeded, against all odds during the Reagan administration, and the cedars were protected. Years later, Bonker's successor in the Third District, Rep. Jolene Unsoeld (D), managed to secure additional appropriations to purchase second-growth buffer lands around the grove to protect it from windthrow.

Now all timber harvest (second-growth federal trees swapped in partial exchange for the cedar grove) has been completed. Jody Atkinson oversaw the logging for the U.S. Fish and Wildlife Service, Weyerhaeuser loggers cooperated, and the damage was minimized. At last the entirety of Long Island is protected within the Willapa Bay National Wildlife Refuge. It is good that these things transpired under Bonker and Unsoeld: their conservative successor, Rep. Linda Smith (R), made it clear that she was no friend of endangered species or federal "set-asides." Smith's own successor, Rep. Brian Baird (D), presided over further critical refuge additions, including prime on-shore elk habitat; but his follow-up act, Rep. Jaime Herrera Beutler (R), has nixed some further refuge expansion. Conservation comes in waves and windows, but thanks to recent gerrymandering of the Third District, this R/D teeter-totter may be stalled for now on the R side. Not that the Republicans have always resisted conservation, as the next stories show.

After local resident Jack Scharbach brought it to their attention, Hendrickson Canyon attracted much interest from state ecologists. We found others eager to protect a heritage forest in these hills, including loggers and their families. I sat on the Commissioner of Public Lands-appointed Natural Heritage Advisory Council for eight years, pestering my colleagues incessantly, as Hendrickson slowly rose from a proposed preserve to the number one candidate for protection under the Trust Lands Transfer program (TLT). An innovative approach, TLT allows trust land, administered by the Department of Natural Resources (DNR) for income to schools and counties, to be transferred into protected status as either a Natural Area Preserve (NAP) or Natural Resource Conservation Area (NRCA). This minefield of acronyms is a small miracle, taking ecologically unique state lands, not the best for logging anyway in most cases, off the rolls of the money-making acres, while losing nothing for the trusts. Tracts are purchased with general funds which then go directly into school accounts, as would normally occur only when the land was logged: a win-win if ever there was one.

But just as Hendrickson Canyon was poised to be transferred, the legislature approved no funds for the program, and we went back to waiting and lobbying for another five years.

An earlier Commissioner of Public Lands, Brian Boyle (R), strongly supported the TLT program and made an administrative withdrawal of Hendrickson to protect it temporarily. The next commissioner to be elected, Jennifer Belcher (D), maintained that status. But this could not go on forever, and there was pressure to make more state timber available for sale. Eventually, formal preserve dedication had to be achieved if we were to rest easy. When the TLT again came up for appropriations, I asked the Wahkiakum County Board of Commissioners, led by Esther Gregg, to pass a resolution in favor of a Hendrickson Canyon NRCA. They came through with key support, as did the Gray's River Grange. Then our local senator, Senate Majority Leader Sid Snyder, a Democrat from Long Beach, gave his vital endorsement. The legislature not only passed the appropriation, but placed Hendrickson Canyon in the top "must acquire" category. But it still had to be approved by the Board of Natural Resources.

As the key meeting of the Board approached, it probably didn't help when, as I lobbied her for the umteenth time, I swept a glass of red wine all over Commissioner Belcher's blue suit, just before her scheduled speech. She forgot and forgave, but stood back next time. Now a new and unexpected problem arose. Because the canyon is home to nesting marbled murrelets, a threatened species that came into prominence after the spotted owl, state appraisers rated its value very low. This meant the University Trust would receive little value for the land and timber. There was a real danger of the Board turning down the transfer on that account. But DNR economists devised a bold scheme to apply substantial economic value for non-commercial habitat. Armed with this new tool, and in spite of last-ditch resistance from antis, Jennifer Belcher muscled it through in one of her last acts as Commissioner of Public Lands. Hendrickson Canyon was finally saved after a campaign of nearly twenty years, on December 5, 2000. Celebrating with supporters at Olympia's Fishbowl Brewpub, I literally cried into my beer. And what a marvelous sequel, when Nirvana bassist/community activist Krist Novoselić bought the quarter section of land immediately adjacent to Hendrickson Canyon, doubling the amount of protected land. Krist and his wife, Darbury, became volunteer stewards for the nature reserve.

That long-awaited victory helped balance the loss of yew-rich Ellis Creek and the rare noble fir stand atop the tallest Willapa hill, Boistfort Peak, both of which Weyerhaeuser logged in spite of conservationists' efforts to protect them. While such losses continued, several old-growth fragments unknown to me when I first wrote *Wintergreen* were located by Washington Natural Heritage Program biologists and others, and some were saved. One of these stands is a steep square-mile in the Gray's River headwaters, part of which became the Willapa Divide NAP.

Well to the west, on the other side of Radar Hill, the biologists found a substantial stand of naturally regenerated hemlocks that grew up following deforestation by the great 'Twenty-one Blow, interspersed with hundreds of giant western red cedars that withstood the storm, and a great old-growth Sitka spruce forest as well. One day, with planner-birder Jim Sayce, I saw the only goshawk I have ever seen in these hills shooting over the deep green crease of the South Nemah River. When then-Commissioner Boyle came to see the incognito, he encountered friends of ours, Sue and Norm Osterman, who took him inside a great hollow cedar we all loved. He was impressed. As a direct result, I believe, he bumped up the site's rating, and some 1,400 acres were secured as the South Nemah Cedars NRCA. Years later, on the same day they created 159-acre Hendrickson Canyon NRCA, the Board of Natural Resources acted to enlarge both Willapa Divide and South Nemah, to 587 and 2,439 acres respectively, safeguarding their big trees, marbled murrelets, and rare salamanders for as long as these hills last. Other small additions have followed from time to time, when TLT funds have become available.

The old growth inventory also includes a pair of upland Sitka spruce forests in Forts Canby and Columbia, now safe as state park natural forests; and a pair of tidewater spruce "surge-plains," protected by DNR, The Nature Conservancy, and the Columbia Land Trust on the lower Chehalis and Gray's rivers. Additional reserves established on the estuaries of the Niawiakum, Bone, and Elk rivers brought vociferous opposition from locals concerned that hunting and public access might suffer. Once these fears were addressed, one or two politicians continued to fight the whole idea, but their objections showed up as petty in light of the tiny acreages involved compared to the vast industrial forest estate. Other local citizens lent their passionate support, and good sense prevailed in the end. Interestingly, one of the local citizens concerned about abridgement of hunting and fishing by NAPs, Rep. Brian Blake (D), got into politics over the issue. Now chairman of the House Agriculture and Natural Resources Committee, he is a close observer of nature.

Two more new reserves give additional cause for celebration, both originating in the vigilance and creative activism of a local logger-turned-filmmaker. Situated hard by U.S. 101 where it rounds Willapa Bay, the Teal Slough giant cedar stand lies within easy reach of several schools. It is frequently visited by college and school groups, tours, and others wishing to experience old growth but unable or to go out to Long Island. But it was very nearly logged.

One man spearheaded the effort to save the stand. Rex Ziak belonged to a four-generation logging family and had worked in the woods himself. Familiar with the timber business, he spotted the flagging that signaled a future timber sale. Determined that the trees should stand, he went to work convincing others. An Emmy-winning cinematographer, Ziak took time out of a busy filming schedule to lobby the owners, at his own expense. Doing everything right that Michael Moore did wrong in *Roger and Me*, he donned a sport jacket, plied secretaries with gifts of canned salmon, and penetrated the upper floors of the John Hancock Tower in Boston. Belling the cats in their executive lair, he sought clemency for the remarkable educational opportunity represented by the little Teal Slough eyebrow above the bay. Before they knew what he was doing, he took out a rope the circumference of the trees Hancock was about to cut, spread it on the boardroom floor, and invited the absentee landlords to step inside. In the end he persuaded, cajoled, and/or shamed John Hancock executrons into sparing the grove for the time being. Eventually, The Nature Conservancy (TNC) stepped in to help the state buy Teal Slough from Hancock for dedication as an NRCA. Now, when we wish to show students a bit of forest as it was and could be; share with visitors a cedar that predates Shakespeare and rivals the redwoods in size; or take the tonic of the deep old wood for ourselves, and haven't time to canoe to Long Island or hike into the South Nemah, we come here.

Rex's influence didn't stop with Teal Slough. Hancock held another hunk of leftover old growth in the misty heights above the Naselle estuary. Most of my previous experience with Ellsworth Slough had come from a couple of memorable outings. The first was a hot, birthday paddle with Thea up the creek, swimming from the canoe and discovering a colony of dun skippers at the forest's marge. On the second occasion, we accompanied a bizarre expedition of young ancient forest advocates who stormed in from Seattle to demonstrate their solidarity with the woods. Bizarre, because their zeal was unmatched by any experience afield. One of them promptly got lost. Our vigil ended with pancakes at the house of friends on Parpala Road and the radioed news that the ardent urban warrior had walked out the long way and hitchhiked home.

Ellsworth is a wonderland of steep ravines, gin-clear streams, massive spruces and cedars, giant sword ferns, and arm-deep sphagnum mats—a republic of Deep Green. Recently, prowling the canyons and flats with Rex Ziak and Cathy Maxwell, I found the first woolly chanterelles I've ever seen in Willapa, and red-breasted nuthatches, rare in the logged-over hills, beeped in abundance. For a long time it seemed as if the place was doomed to certain logging, and indeed parts of it were carved away. Roads and landings stood ready to take the rest. But in view of the attention given in *Orion Afield* magazine and elsewhere to Ziak's earlier campaign, the company decided it would rather not attract the harsh PR that cutting this site would bring, and they offered it for sale. The Nature Conservancy announced plans to buy and restore the entire 5,000+ acre watershed of Ellsworth Creek, assembling the whole from the remaining old growth acres cobbled together with diverse, older second growth and some recently logged lands. The DNR abetted the effort by establishing an NRCA on adjacent Elkhorn Creek. Not long into the campaign, the Paul Allen Foundation kicked in two and a half million dollars to get it going—the largest contribution TNC's state office had ever received. The full-time manager, Tom Kollasch, has a challenging job balancing restoration with an unruly pack of opinions about how the site should be managed to serve the community as well as the creatures. TNC has since pulled out of its smaller holdings in southwest Washington, but retains Ellsworth as a showcase landscape-level project. All this in little old, post-old-growth Willapa—unimaginable a few years ago! Now it seems that Ellsworth *and* Teal Slough's forests will continue to stand, to signify that big old trees and their communities still matter in the wintergreen land.

Lest it seem as if the land is all being locked up, as the zanier antis contend, let's place it in perspective: *The remains of the ancient forest of Willapa amount to well under ten square miles, out of some 2,500 square miles formerly graced by one of the greatest woodlands on Earth.* Anyone who contends that the conservationists have been grabby had better be good at keeping a straight face. It may not be much, but the fact that we're in better shape on old growth than I thought we were in 1986 has come as a pleasant surprise. An even larger surprise has been the rapid liquidation of the second growth.

When I came to Gray's River in 1978, I found much naturally seeded second-growth forest, largely western hemlock with some Douglas-fir, western red cedar, and other species. Some of it had been planted, but without the slash burning and herbicide spraying that are standard practice today, and not in rows of genetically selected, engineered, or cloned supertrees. The trees became big and nicely spaced; the understory not too tangled, yet relatively wild and growing in diversity by the year. There were lots of chanterelles, torrent salamanders, and other life forms intolerant of freshly disturbed, over-young, or simplified monoculture woodlots. In short, these were woods on the way to becoming real forests. We took a lot of pleasure in them, and yet we also took them for granted, because they were not old growth. How could we have known that vast tracts of the good secondary woods would be liquidated in between editions of this book—literally shaven in fifteen years' time? Woods that could and should have provided jobs, fiber, and diversity for fifty years plus.

The sack of the second growth came about for several reasons. The unions were largely gone from the woods, and the incentive was therefore raised for independents and contractors to cut as much and as fast as possible to pay their bills. There was a rush to export timber at peak prices before state and federal restrictions on exporting raw logs came into effect. Reduction of federal acres open to timber sales due to President Clinton's Northwest Forest Plan put a premium on private stumpage, encouraging gyppos to scour the hills for every widow's back forty or front-yard five. A new influx of settlers and developers raised land prices for homesites, inspiring some owners to barber the trees and bulldoze the stands for quick conversion to real estate. But the most important element was the transfer of enormous acreage from old timber concerns to general investors and holding companies.

Weyerhaeuser and Crown Zellerbach companies catch some severe criticism in *Wintergreen*. Yet, for all their rough handling of the woods and the people of the woods, they at least seemed devoted to forestry in the long term back then. But as Weyco converted more and more land to commercial real estate, I wondered. Then Weyerhaeuser, in rationalizing its holdings, sold thousands of acres in Willapa to the John Hancock insurance giant. Crown Zellerbach was snapped up by British junk-bonds investor Sir James Goldsmith and later converted to Cavenham Forest Industries under the U.K.-U.S. firm, Hanson plc. Neither Hancock nor Hanson had any history of commitment to sustainable forests; shareholders' profits were their sole concern. With decisions now coming not from Tacoma and Portland but Boston and London, these entities proceeded to log the second growth at rates undreamed of by the former landlords. Crown's former Cathlamet Managed Forest has since changed hands at least twice more. My visionary neighbor Bobby Larson sagely proposed that the county should acquire the whole vast estate, thus bringing its future back home; but this was never pursued. Had it happened, our income-starved little county would have a steady flow of timber receipts into the budget, instead of waiting every year to hear how fluctuating stumpage prices and diminishing state timber sales will dictate economies and cuts. At least a state bill was passed, pushed hard by Wahkiakum County Commissioner Dan Cothren, requiring timber-rich counties to share receipts with timber-strapped counties.

Now one wonders if Weyerhaeuser is having second thoughts about trimming its kingdom, having recently purchased 645,000 million acres from Longview Timber. But however much the principals in the woods have played musical chairs, the outcome was the same: before we could turn around, the chanterelle woods were gone. In their place, vast expanses of dense young doghair hemlock and attempted Douglas-fir plantations covered the slopes. Willapa is still wintergreen, but the overstory often looks like undergrowth.

Nor is the horizon likely to change back into big trees. The American and Japanese appetites for cellulose pulp have converted much of the Columbia River lowlands from Skamokawa to Puget Island to Kalama into hybrid cottonwood plantations scheduled for cutting every seven years, and the uplands into hemlock thickets that will be lucky to stand for thirty. Friends in the mills tell me not to look for many sawlogs being grown any more, but to expect chipboard and other composition products made from smallwood fiber to take over. That means shorter and shorter cycles of rotation—sticks barely old enough to vote now go down the road on the logging trucks in bundles smaller than single trees of yore. Skinnier and skinnier trees are going to make up the forests of tomorrow. And when the reprod is all doghair, the diversity goes to pot.

State timberlands, the only public forests in Willapa and not much compared to the private holdings, have been managed somewhat more conservatively. Yet in spite of grumbling over new regulations, DNR field personnel and contract loggers continue to cruise and harvest plenty of board feet. The September 14, 1995 *Wahkiakum County Eagle* announced that "Conservation pressure hampers DNR harvest," and went on to detail officials' concerns over the rules. Yet two weeks later, the *Eagle* reported: "State trust lands expected to produce highest revenue in history for next six years." This level of cutting is good for taxpayers, but whether the forest can sustain it remains to be seen. Forest practice regulations have been slightly stiffened, and "new forestry" leaves a few standing live trees and snags behind as well as woody debris on the ground. But these "reforms" often lead to self-parody: "green-tree retention" amounts to two trees per acre in one current harvest scheme. Examples of bad logging still abound. The gouged-out road and take-some, leave-some logging committed on the once-beautiful forested knoll above Naselle School looks like the most bizarre of the kids' sculptured buzzcut hairdos: uglier than sin. Enormous, old-style clearcuts are not yet merely memories. Nor are jokes of streambank buffers, one alder deep. Smaller streams, Class IV and V in the state rating system based on anadromous fish, still require no setbacks at all, and I have seen recent clearcuts right down to the shoreline of the Class I Naselle River. Thinly stretched enforcement and paltry fines, easily absorbed as the cost of doing business, encourage some operators to ignore even mild new initiatives toward better forestry. Soil still runs into the salmon streams, and the herbicides still make bitter rain in spring.

As I wrote in "The Sack of the Woods," Willapa being essentially post old-growth, the northern spotted owl is extinct here, so the furor around its protection didn't affect things here much, as in the Cascades. (In any case, the adaptable replacement species, the barred owl, is abundant here now.) And, these being all private or state timberlands, President Clinton's Forest Plan doesn't touch them much either. But one other element in recent years has touched Willapa. That was the listing of the marbled murrelet as a threatened species. Murrelets nest on broad, mossy old-growth boughs, and a few birds still commute from Willapa to the sea and back nightly. So a few parcels have been withdrawn from logging for murrelet protection, causing cash-strapped commissioners to fulminate against them (never mind their total habitat would take loggers only days to liquidate). A stroke of bad luck landed the best site for wind power production, on Rader Ridge, immediately adjacent to a major Marbled Murrelet Management Area (MOMA) overlapping the South Nemah Cedars NRCA. PUD and county commissioners trying to comply with a green power mandate from Olympia pushed the wind farm, only to find the conservationists agin 'em because of marbeled murrelets. I understood their frustration, but the windmills would have killed some number of the very rare seabirds, and the project had to die a'borning.

Against such business as usual and unusual, a remarkable development occurred that briefly cast a brighter light on the misty hills and their ring of rivers and bays. My former forestry school classmate and colleague in The Nature Conservancy, Spencer Beebe, saw the need for a) protection of the temperate rainforests, and b) homegrown, locally controlled institutions for planning ecologically sound, sustainable development. He left Conservation International, his former brain child, to found Ecotrust, a Portland-based organization dedicated to those premises. One of their early moves, in conjunction with the Washington Office of The Nature Conservancy, was to catalyze a local consortium known as the Willapa Alliance in 1992. The Alliance sought to rally oyster, cranberry, timber, fish, tourism, and other conservation-minded folk to work for a sustainable economy in the Willapa watershed. The Alliance also tried to marshal forces to combat invasive *Spartina* cordgrass in Willapa Bay, while spurring publication of an impressive book (*A Tidewater Place*, by Edward Wolf) and a CD-ROM interpreting the watershed (*Understanding Willapa*, Good Northwest/Pacific GIS).

Most importantly, the Alliance got everyone talking (not always comfortably) about resource issues that concern every person and every species here. As former Alliance Director Dan'l Markham said, "The Alliance provides a forum for gaining new insight into the workings of the Willapa ecosystem and for finding creative ways to ensure the longterm well-being of Willapa's communities, lands, and waters" through science, management, education, involvement, and sustainable development. A list of social and biological indicators of regional health and sustainability was prepared, to guide the group's further growth and programs. The timber giants were invited to participate, in hopes of developing some model forestry and eco-assessment techniques. Weyerhaeuser invested a half-million dollars in the ShoreTrust Trading Group, a lending institution started by the Alliance to fund ecologically viable projects. Cavenham/Hanson put $20,000 into local emergency services. Hancock Timber Resources Group cooperated as a business-like but willing seller of some estuarine reserve lands.

But the problems inhering in a watershed obliged to satisfy numerous conflicting expectations are many and formidable. Cranberries crashed when the huge Ocean Spray cooperative failed to support smaller producers. Traditional oyster growers continued to spray their beds with Sevin (carbaryl) to kill mud-shrimp, and became embattled with proponents of an unpolluted Willapa Bay. (The Sevin was eventually phased out, but now they want to use a neonicotinoid pesticide (imidacloprid), which can be very dangerous to non-target invertebrates. When toxic chemicals must be employed to sustain a non-native species (Japanese oysters) against a native one (mud shrimp) in a so-called pristine estuary, couldn't there be a flaw in the picture?) Though the neonics are a recent concern after big bee die-offs, such questions plagued the Alliance from the start.

A parallel debate played out over introduced East Coast cordgrass *Spartina alternifolia*, which was rapidly converting productive mud flats to pretty but biologically simplified meadows. *Spartina*-fighting agencies fought bitterly with organic oyster growers and other citizens opposed to spraying the bay with Rodeo. I was conflicted: while agreeing *Spartina* altered the estuary's ecology, I was concerned that no one had documented the dangers of glyphosate to the phytoplankton or the bay as a whole. A plant-sucking leafhopper recruited and tested as a biocontrol agent buoyed hopes, and I had my own hope for mechanical control. But the chemical forces prevailed, aggressive spraying proceeded, and cordgrass has been impressively reduced—at what cost to the eco-system, we cannot yet know. Meanwhile, some oystermen want to begin spraying Japanese eelgrass, hard to tell from the essential native eelgrass; the bay has been invaded by exotic green crabs, and native Dungeness crabs are under threat from Army Corps plans to dredge the Columbia deeper and dump the spoils on off-shore crab beds, even as the Pacific removes the beaches of Outer Willapa faster than they can be laid down. And so it goes.

The Willapa Alliance was a good dream. But like many dreams, it grew chaotic, and eventually dissipated like mists over the oyster beds when everyone woke up to their ongoing, inexorable conflicts. Some poor staffing choices and city/country language barriers didn't help. Maybe the job was just too big; maybe the hopes too high. Or maybe the challenges the Alliance undertook are simply intractable in a time when philistine mercantilism rules the day and population pressures, globalism, rural decay, and climate change, fueled by suspicion and ideology, make for constantly shifting ground and rising water. Eventually Ecotrust pulled back to Portland, leaving a vital presence in the form of an important bioregional web source based in Gray's River (www.tidepool.org), run by writer Ed Hunt, for a while; and an alternative, community financial resource, ShoreBank, until some big ordinary bank bought it. And the Alliance faded away. Anyway, along came salmon, overwhelming everything else.

Not that salmon are anything new here. They were, after all, one of the three great legs of the economy that built the place, along with logs and milk. While dairying has largely left and logging has metamorphosed into pulp farming, the fishery too went to hell. Between dams on the Columbia, overfishing, degradation of spawning beds from upstream road-building and steep-slope logging, herbicides and industrial runoff, and other factors, the great fish haven't had a chance. The federal listing of eleven runs of Northwest salmonids as threatened or endangered species under the Clinton regime sought to turn that around. Native fishing rights (complicated by the rise of reservation casinos and federal recognition of the Chinook Tribe, accomplished under Clinton and immediately retracted by G. W Bush), the future of gillnetting on the Columbia as a way of life, the role of hatcheries as policy favoring genetically native stocks replaces the old mix'n'match philosophy, and a thousand other issues will all be militated by Grandfather Salmon. Grandmother Salmon's First Harvest may even come back to waters long deprived, if blocking dams on the Snake and other rivers actually come down, as they already have on the Elwha and White Salmon. Monthly, one reads of new salmon plans, compacts, consortiums, and commissions. The Columbia deepening was held up by salmon. Caspian terns, elegant black-capped fishers with fire-engine-red bills, as well as double-crested cormorants, have become public enemies, because they nest on dredge-spoil islands in unprecedented numbers—and eat salmon. Marine mammals pit their protectors against those who feel sea lions eat too many fish. Water-quality decisions are driven by salmon as much as people. Everything that happens from here on out will have salmon scales all over it.

I earnestly hope that the infinitely complex brokerings and tinkerings to come will somehow restore the viability of the Columbia gillnet fishery. This may be asking too much. The old tension between sport fishers and commercial fishers has come to a head with the governors of both Washington and Oregon bowing to their fish and wildlife commissions' desires to favor the more lucrative sport lobby and all but kill off the traditional, sustainable, and culturally and economically vital gillnetting industry on the main stem of the Columbia. The architects of this plan say it's for conservation, but I believe that is them being cynical: they just prefer thousands of fishing licenses and boat-and-gear-sales taxes over the families—30-40 in Wahkiakum alone—who still depend upon commercial fishing.

One only needs to read Irene Martin's elegant history of Wahkiakum County, *Beach of Heaven*, and her other books drawn from that rich culture, to see what gillnetting has meant to the human texture of the place. Irene and her fisherman husband, Kent, have toiled and agitated in the salmon battles with an intelligent and informed vigor that activists who love their place anywhere on earth might well emulate. Another couple, Cathy Maxwell and her late husband Ed, have done the same from different vantages—Ed as the supervisor of salmon hatcheries in the region, Cathy as the consummate plantswoman. In the process, people like the Martins and Maxwells have taught me much of what I know about this place.

For it is the people of Willapa who have been *Wintergreen*'s greatest gift to me. For an upstart outsider to be able to come in and comment, not always gently, on a rich and settled community's way of life, and to be tolerated, is a mark of a good people in a good place. There have been a few threats and jake brakes from those who never read the book and assumed the worst about me; but there have also been loggers in red suspenders and tin pants knocking at my back door, asking to buy the book, and would I sign it? Sadly, much of the community's human old growth has passed in my time here. Bob Torppa, Jim Fauver, Ed Sorenson, Walt and Mary Kandoll, Harold Badger, Johnny Kapron, Norman Durrah, Veryl and Barbara Chamberlain, Helen King, Opal Kraft, Norman and Myrtle Anderson, Jenny Pearson, Glenrose Hedlund, and postmaster Jean Calhoun have all gone on, and most recently, my oldest friend in Gray's River, Marilyn Gudmundsen. Yet Bobby Larson still holds forth at Grange, and Carlton Appelo at the telephone company. New folks come in, some of them influenced by the book; my friend and neighbor Steve Puddicombe even named his pretty spread, the old Kandoll place on the banks of Gray's River, "Wintergreen Farm."

Rex Ziak of Naselle is one of the delights among the people I've come to know since I wrote the first edition. Krist Novoselić, who became and remains master of Gray's River Grange, is another. Then there is Walkin' James Powell. One day I got a letter from a boy across the hills in Menlo. Soon thereafter, he *walked* the ragged way from Menlo to Gray's River, thirty miles across the hills on logging roads, to visit us. He had read *Wintergreen* and felt he'd found a soulmate . . . not much of a fieldmate, however: James routinely walked dozens of miles a day across rough country, and I never could keep up with him. But it's been a marvel to watch this naturalist prodigy, who began with birds and big trees and moved on to plants and the rest, develop into a grown and educated resource professional who has mapped the land for Ecotrust, pruned trees for future forests, and gobbled the Willapa with the walker's appetite, naturalist's hunger, and lanky stride, like some Bob Marshall of the clearcuts instead of wilderness.

We know so much more now about the natural history of the Willapa Hills. Alan Richards, Ann Musché, Andrew Emlen and others have traced in the birds nicely, Jim Atkinson the herpetofauna. Ed Maxwell found more western toads and tailed frogs along the Naselle River, though in small numbers; and Thea, with Bruce and Terry Satterlund, finally discovered masses of toads breeding almost in our backyard, on the West Fork . . . after all these years of searching all over the Hills! Cathy Maxwell has recorded many more plants in the hills, including some first state records and some disjuncts from the Olympic Mountains and Oregon Coast Range, and published her opus: "Vascular Flora of the Willapa Hills and Lower Columbia River Area of Southwest Washington" (*Douglasia Occasional Papers* iv: 1991). We have gone from next to no knowledge about our butterflies to a known fauna of some fifty species, many found in our own butterfly garden, ten percent of which were here today. The subtle and often-ignored aspect of Willapa has its advantages in rendering discovery easy to come by. For example, no one has ever attended much to the moths and dragonflies here; Lars Krabo and Dennis Paulson, respectively their leading Northwest authorities, are helping us to flesh out lists. The highly diverse flies, the waterbears, the lichens, and such await their chroniclers. Meanwhile, the flora and fauna don't sit still: elk are still caught in nets and transferred out of the Columbian white-tailed deer's reserve, now known as the Julia Butler Hansen National Wildlife Refuge. Bears have been on the increase, since the people of Washington passed an initiative aimed at ending the hunting of bears with hounds and baits; but I have yet to see one here.

If nature's close study remains the domain of the few, conservation has become everybody's business. When I wrote *Wintergreen*, except for reports on timber harvest levels, resource stories were rare in the local press. Now you can't pick up a paper without reading headlines such as these, hot off the press as I write: "Senate Democrats seek to protect rural areas from Shoreline regulations" (*Wahkiakum County Eagle*), "Crucial habitat project adds 871 Chinook acres" (*Chinook Observer*), and "Fishermen, environmentalists sue EPA over pesticides and salmon" (*The Daily Astorian*). Among the heat and bluster, real advances take place. For example, the Columbia Land Trust has used salmon recovery money to purchase and restore several major floodplain and abandoned farmland habitats, though not without ruction among those who think they are gobbling up the county. The Grays River Habitat Enhancement District and a dozen other agencies and programs try to bend the river to their will, and the river flows on regardless, and pretty much does what it wants.

Historical preservation has also made gains. Across K-M Mountain from here, in the river-village of Skamokawa, a handsome old wooden belltower looms out of the "smoke on the waters," which is what Skamokawa means in Chinookan. Once a schoolhouse, then a lodge,

Redmen Hall has been restored by Friends of Skamokawa into a fine Riverlife Interpretive Center and a locus for music and the arts. Across Brooks Slough, Silverman's Emporium once served steamboats calling at the town. Its pilings rotting, the venerable riverside mercantile seemed doomed to slide into the Columbia. But it was found, rescued, and carefully restored by Arnold Andersen, who made its beautiful Lurline Auditorium available for annual community shows and auctions; now, offices and a B&B. The county's first condos mounted an invasion on the mouth of Skamokawa Creek, but the project failed, leaving a renovated village shop, a popular B&B, and a kayak center in its wake. Historical churches have been restored in Cathlamet and up Deep River, and the stately Cathlamet Hotel is in business again. The much-beloved Gray's River Covered Bridge has been rebuilt to its old plan (except strong enough now for the school bus and milk truck), and for several years had its own festival each August, when the county's population would double for a day. Too much work for too few folks, this became a simple picnic, and then an annual 4-H dinner inside the bridge; all ways to honor the grand old structure. The Grange's Ahlberg Park beside the bridge, formerly part of Bobby Larson's dairy farm, welcomes all visitors and honors H. P. Ahlberg, who built my house, started the Grange, and then went back to Sweden, leaving his daughter Ebba and the Sorenson family she married into to carry on Swede Park.

What with the dairy herds nearly all shipped out and the post office stolen off to Rosburg, our village has fallen into an even deeper torpor, not to say that it's moribund. One fine dairy remains, the Burkhalters' of Rosburg, certified to milk for Organic Valley. The cafe and the store are temporarily closed again, but Duffy's Irish Pub thrives in Gray's River, having arisen from the ruins of the long-dead Valley Tavern. Restarauteur Al Salazar has created an almost baroque collection of funky and charming buildings alongside the riverbank, and bartender-without-peer Lorraine serves a perfect Guinness or Dick's IPA from Chehalis for sipping on the porch among the swallows and bald eagles.

Only twenty miles to the east, the redoubtable River Mile 38 Brewpub has materialized beside the Cathlamet Marina, serving superb local ales. Now this one never would have guessed or believed, back in the early *Wintergreen* days of nothing to drink but Rainier Ale! And fifteen miles west in Pacific County, a fine branch of the Timberland Regional Library arose close enough to make its beneficent presence felt, after the good folk of Wahkiakum County twice turned down a library district initiative by a few votes. For all the gains and losses, we still have no traffic lights in the entire county, deer and elk still outnumber people except during hunting season, and the night skies are still dark.

Realizing they could do better raising trailers than trees, a few folks have cleared chunks of forest for homesites, and manufactured homes and a few stick-built houses have sprouted here and there on scraped-off benches above the river. Fair enough. For a while, an unsuitable subdivision planned for the floodplain threatened to change the valley significantly. But after a struggle, cooler heads prevailed, and a better future for everyone came about. Now those fields are thick with red clover and orange sulfur butterflies in high summer, and beautiful Wagyu cattle the year round.

For now, unimpressed by people's plans for the valley, the elk walk where they will, although of late they are limping, thanks to a troubling epidemic of foot-rot. The river still rises and falls to the venting of the clouds, the petitioning of the salmon. As predicted in an earlier edition of this book, Captain Robert Gray and crew indeed came back to the Gray's River tidal basin on May 15, 1992, the bicentennial of his crossing of the bar in the *Columbia Rediviva*. Lots of people came after all, including schoolchildren, folksingers, and Chinook Indians beating drums in Plains Indian headdresses. Grangers in period boats and dress approximated Capt. Gray and mates, rowed up the river by the local basketball team not quite in period dress but still wearing the pioneers' Nordic names. Wirkkalas and Penttilas, Johnsons and Ericksons were much in evidence. Old families, and some of the old ways, persist among the new. Next, the Lower Columbia region braced for the bicentennial of the coming of Lewis & Clark, who spent the wettest, most uncomfortable nights of their entire trip at a place near here they called the "Dismal Nitch," at the onset of the autumn rains of 1805. That was a five-year orgy of remembrance and pilgrimage, which left us (thanks not in small part to Rex Ziak) with a new national park, lots of tourist dollars, and the hope that California condors, such as Clark and Lewis found here, might one day come back too.

For now, even in Willapa, comes the time of the three C's: carbon, climate, and catastrophe. Concerning carbon, we lie happily outside anyone's idea of a good place to frack or drill. But we have just lived through a citizen's Battle Royale to beat back a plan most foul, to install a liquified natural gas terminal across the Columbia from Puget Island. After the speculators left town, among their debts and detritus was found a list of the places that would be incinerated in case of an accident, in order. The home and farm of some dear friends of mine were # 8 on the list. Need I say more? Two other LNG proposals survive for now, but I have faith that they will go down as well. Meanwhile, the folks of Longview, in southeast Willapa, and Grays Harbor in the northwest, are fighting off coal and propane and oil trains and terminals right and left.

As for climate, wood nymphs seem to have colonized western Wahkiakum County for good, and the five kinds of butterflies out there today in March tell the same tale, as does frogsong hitting its full pitch in January, daffodils outpacing crocuses, and various other phenological disconformities such as this very spring has exemplified: we too will experience the sharp edge of change, the great warming and drying that will render many places harder and harder to occupy. We may get the easier end of things to come in this green and pleasant corner; but even that may bring change, if we are finally discovered, heaven forbid, as a viable place to live, as the rest of the continent withers, freezes, or just blows away.

But the third coming thing is one from which we shall have no escape, at least on our coastal margins and sea-level invaginations: catastrophe, and its aftermath. For the sea level is rising, and the Cascadean subduction event and its attendant tsunami are on their way, already overdue. The Juan de Fuca Plate is slipping beneath the North American Plate, and is presently hung up. When they crack free, in a nine-point earthquake, the coast will drop several feet, the great wave will follow, and our shores and valleys, peninsula and estuaries, flats and floodplains, hollows and homes, will become very different from the way we know them, very fast. We can all hope it holds off, and if the crack-up comes up or down the coast we may not get the worst of it. But the Great Adjustment will surely come in the next century or two (current estimates give it 17-35% chance in the next fifty years) and the bones of the land will roll again in the rapture of orogeny.

Meanwhile, my old home place, Swede Park, has stood here for more than half the time since Lewis and Clark came. It perches above the registered hundred-year farm of the Ahlbergs and Sorensons, returning now to marsh and willow-brake. One Swede, Thea Linnaea, lived here until she passed way too damn soon in late 2013. Our honorary Finn, the great cat Bokis Volkilla, lasted nearly twenty years, and resides beneath the plum grove where fringe-cup blooms in a wave of fragrance each May.

Now one Anglo-German writer with a long-ago tincture of Algonquin lives here still, with a shelter-cat named Bo Diddly who keeps the Steller's jays on their toes but mostly bedevils the garden voles. European maples, English oaks and ivy, and Armenian blackberries compete with native alders and spruces to see who will finally overwhelm the place. Across the valley, the Timbered Tor is timbered again. Both frosts and floods became more frequent and severe for a while, but have backed off again. And even though we have just endured the sunniest late winter in forty years or more, the long range forecast is still for rain.

*Gray's River*
*March 18, 2015*

# ACKNOWLEDGMENTS

*Wintergreen* is the product of rubbing up against the land of Willapa and its people. I wish to thank the people of Gray's River and Wahkiakum and Pacific counties, Washington, for their warmth and hospitality and for the education they have given me. In particular I am owing to Kent and Irene Martin, Ed and Cathy Maxwell, Jean Calhoun, Carlton Appelo, Marilyn Gudmundsen, Ann Musché and Alan Richards, Veryl Chamberlain, the Joel Fitts family, Ed and Lenore Sorenson, Marilyn Gudmundsen, Joe Florek, the Gary and Carol Ervest family, the Mike and Diane Matthews family, Steve McClain, Sunrise and Jessica Fletcher, Red Almer, Bob Torppa, the Robert Larson family, David and Elaine Myers, André Stepankowsky, Marc Hudson and Helen Mundy, Bob Richards, Irene and Steve Bachhuber, Dick Moulton, Dennis Stein, Toni Scott, Rick Nelson, Carol Carver and George Exum, John Schmand, Jack Scharbach, Just Thomas, Bob LiaBraaten, and Walter Kandoll for talking about Willapa with me. Many others, newcomers, old-timers, and outsiders alike, named or not, will recognize their contributions.

Resource-agency people including Susan Saul, Jody and Jim Atkinson, Jim Hidy, Jerry Franklin, Gary Hagedorn, David Hoffman, Jay Brightbill, Elizabeth Rodrick, Mark Sheehan, and Dennis Nagasawa shared their knowledge of the hills and forests and their wildlife with me. David McCorkle, Paul Hammond, Charles Remington, Rod

Crawford, Kris Schoyen, Ralph Widrig, Vance Tartar, Ingrith Deyrup Olsen, Grant W. Sharpe, Nalini Nadkarni, Kathleen Sayce, Jim Sayce, Dick Wilson, and Louis LaPierre have all provided valuable biological information about the area.

Since the first edition, my intimacy with these hills has increased thanks to Rex Ziak, James Powell, Mark Scott, Lorne Wirkkala, Hans and Jenelle Varila, Kyle Matthews, Ed Hunt, Steve Puddicombe, Mary Steller, Bryan Pentilla, Krist and Darbury Novoselić, Merlin and Judy Durrah, Dan Cothren, Ted Wolf, Andrew Emlen, Mike Patterson, Bruce and Terri Satterlund. and many others. Special thanks to all the Zimmermans of Gray's River Valley for all their kindnesses.

Dennis and Kathy Gillespie first brought me to Wahkiakum County. Sally Hughes first came here with me and made it work for us to settle. MaVynee Betsch's generosity made it possible to remain. Thea Pyle saw it out for the long run. Margaret Ford, Fayette Krause, Rod Crawford, Dennis Paulson, and Jon Pelham carefully read portions of the manuscript. Cathy Macdonald very kindly prepared the map of the region. Thanks to Melinda Mueller, David Wagoner, Gary Snyder, and Bruce Springsteen for excerpts of their poems on pages 9, 171, 199, and 333.

First publication of *Wintergreen* owed debts to my original agent, Barbara Williams, and Paul Trachtman, Roger Swain, and Michael Pietsch; Ruth Singleton of Charles Scribner's Sons shepherded the book to print, and I wish upon all writers such a truly helpful editor. *Wintergreen* owes its earlier paperback life to Harry Foster at Houghton Mifflin, to my former agent Jennifer McDonald, and to Gary Luke at Sasquatch Books.

I am immensely grateful to David Guterson for his faith in this book and for championing it from the beginning. For bringing this 30th anniversary edition to press from Pharos Editions, I again thank David for selecting it, the extraordinary publisher Harry Kirchner, and my kind and percipient agent, Laura Blake Peterson, of Curtis Brown, Ltd.

Thea Linnaea Pyle read and improved the entire manuscript in every draft, made the title drawing of single-flowered wintergreens, and for thirty years gave me loving care and aid of every kind that made a writing life at Swede Park work. She is inexpressibly missed around here, and all over Willapa and beyond. Most of all I thank Thea, Tom, Dory, Bokis (and subsequent cats) for living in and exploring the wintergreen land with me.

**Robert Michael Pyle** is the author of eighteen books, including *Chasing Monarchs, The Thunder Tree: Lessons from an Urban Wildland, Sky Time in Gray's River: Living for Keeps in a Forgotten Place*, and the recent poetry collection *Evolution of the Genus Iris*. A Yale-trained ecologist and a Guggenheim fellow, he is a full-time writer and naturalist living in the Willapa Hills of Southwest Washington.

**David Guterson** is the author of the novels *East of the Mountains, Our Lady of the Forest, The Other, Ed King*, and *Snow Falling on Cedars*, which won the PEN/Faulkner Award; two story collections, *The Country Ahead of Us, the Country Behind* and *Problems with People*; a poetry collection, *Songs for a Summons*; a memoir, *Descent*; and *Family Matters: Why Homeschooling Makes Sense*. He lives with his family on Bainbridge Island in Washington State.

## Books by Robert Michael Pyle

*Wintergreen: Rambles in a Ravaged Land*

*The Thunder Tree: Lessons from an Urban Wildland*

*Where Bigfoot Walks: Crossing the Dark Divide*

*Nabokov's Butterflies* (Editor, with Brian Boyd and Dmitri Nabokov)

*Chasing Monarchs: Migrating with the Butterflies of Passage*

*Walking the High Ridge: Life as Field Trip*

*Sky Time in Gray's River: Living for Keeps in a Forgotten Place*

*Mariposa Road: The First Butterfly Big Year*

*Letting the Flies Out* (chapbook: poems, essays, stories)

*The Tangled Bank: Essays from Orion*

*Evolution of the Genus Iris: Poems*

**On Entomology**

*Watching Washington Butterflies*

*The Audubon Society Field Guide to North American Butterflies*

*The IUCN Invertebrate Red Data Book* (with S. M. Wells and N. M. Collins)

*Handbook for Butterfly Watchers*

*Butterflies: A Peterson Color-In Book* (with Roger Tory Peterson and Sarah Anne Hughes)

*Insects: A Peterson Field Guide Coloring Book* (with Kristin Kest)

*The Butterflies of Cascadia*

**More Titles From Pharos Editions**

*Still Life with Insects* by Brian Kiteley
Selected and Introduced by Leah Hager Cohen

*Doctor Glas* by Hjalmar Söderberg, Translated by Rochelle Wright
Selected and Introduced by Tom Rachman

*A German Picturesque* by Jason Schwartz
Selected and Introduced by Ben Marcus

*The Dead Girl* by Melanie Thernstrom
Selected and Introduced by David Shields

*Reapers of the Dust: A Prairie Chronicle* by Lois Phillips Hudson
Selected and Introduced by David Guterson

*The Lists of the Past* by Julie Hayden
Selected and Introduced by Cheryl Strayed

*The Tattooed Heart* & *My Name Is Rose* by Theodora Keogh
Selected and Introduced by Lidia Yuknavitch

*Total Loss Farm: A Year in the Life* by Raymond Mungo
Selected and Introduced by Dana Spiotta

*Crazy Weather* by Charles L. McNichols
Selected and Introduced by Ursula K. Le Guin

*Inside Moves* by Todd Walton
Selected and Introduced by Sherman Alexie

*McTeague: A Story of San Francisco* by Frank Norris
Selected and Introduced by Jonathan Evison

*You Play the Black and the Red Comes Up* by Richard Hallas
Selected and Introduced by Matt Groening

*The Land of Plenty* by Robert Cantwell
Selected and Introduced by Jess Walter

Printed in the United States
by Baker & Taylor Publisher Services